2/06

DISCARDED

The Great Lead Water Pipe Disaster

The Great Lead Water Pipe Disaster

Werner Troesken

The MIT Press
Cambridge, Massachusetts
London, England

MIT Press books may be purchased at special quantity discounts for business or sales promotional use. For information, please email special_sales@mitpress .mit.edu or write to Special Sales Department, The MIT Press, 55 Hayward Street, Cambridge, MA 02142.

This book was set in Sabon on 3B2 by Asco Typesetters, Hong Kong and was printed and bound in the United States of America.

Library of Congress Cataloging-in-Publication Data

Troesken, Werner, 1963–
The great lead water pipe disaster / Werner Troesken.
   p. cm.
Includes bibliographical references and index.
ISBN-13: 978-0-262-20167-4 (alk. paper)
ISBN-10: 0-262-20167-4 (alk. paper)
1. Water-pipes—Materials—History. 2. Lead—Environmental aspects—
History. 3. Lead—Toxicology—History. 4. Drinking water—Lead content—
History. 5. Pipe, Lead—History. 6. Lead poisoning—History. I. Title.

TD491.T67   2007
363.738'492—dc22   2006049401

10 9 8 7 6 5 4 3 2 1

For
Colin Troesken

Contents

Acknowledgments

For reading and commenting on the entire manuscript, I thank Lou Cain, Alex Field, Price Fishback, and several anonymous referees for MIT Press. For commenting on specific chapters or related papers, I thank Dora Costa, Joe Ferrie, Lara Shore Sheppard, Rick Steckel, Tara Watson, and Jessica Wolpaw Reyes. Linton Traub provided detailed comments and suggestions on chapter 2, and on the biochemistry of lead poisoning. Leonard Casson helped with topics related to environmental engineering. I also wish to thank seminar and conference participants at the following venues: the Woodrow Wilson School at Princeton; the National Bureau of Economic Research, particularly Cohort Studies and the DAE group; the Columbia Economic History Workshop; the Indiana University Economic History Workshop; Williams College; the Department of History at the University of Pittsburgh; the Pitt/CMU Applied Micro Workshop; and the Washington DC-area Economic History Workshop. This project was funded by a grant from the National Institutes of Health, P01 AG10120. In addition, I thank the editors at MIT Press for their enthusiasm and attention to detail.

Other individuals who deserve acknowledgment include the following: Paula Canavase and Claudia Linares provided able research assistance at an early stage of this project; Susan Jones and A. J. Aiseirithe made numerous editorial suggestions; Dr. Herbert Needleman shared his many important articles on lead poisoning with me; Robert Fogel first drew my attention to the practice of using lead water pipes; Anthony Bellia of the Notre Dame Law School offered advice on the legal issues surrounding water lead; and Nancy Tannery helped track down useful information on lead poisoning. As with all my work, I am particularly grateful to my wife and colleague Patty Beeson for her encouragement, insights, and suggestions. Of course, I retain sole responsibility for all errors of fact and interpretation that might remain.

Prologue
Exhuming Michael Galler

Late one night in New York City in June 1868, Dr. Marlin Dupuis was summoned to the home of Michael Galler. Galler was vomiting thick, darkly colored blood. Dupuis prescribed diluted sulphuric acid to kill the pathogen, whatever it was, that resided in his patient's gut. The doctor also prescribed a morphine elixir to help ease Galler's nerves. The next day Galler started to feel better, and, while the vomiting continued, it contained no blood. But two days later Galler felt exhausted and observed blood in his stools. Dupuis returned to Galler's home and this time prescribed ammonia along with a mixture of water, sulphuric acid, and champagne. After nearly two weeks, this course of treatment appeared to have a beneficial effect: Galler's mysterious illness disappeared, or at least the symptoms did.[1]

Then, three to four weeks later, in July 1868, Galler had a relapse. He was again vomiting blood, his pulse was feeble, and he was running a high fever. Although Galler said he was not in any pain, Dr. Dupuis prescribed an opium elixir to be taken every two hours. Galler died sometime in early August, reportedly from a bleeding ulcer. A short time later, Galler's widow, Elizabeth, had him buried at the Lutheran Cemetery on Long Island.[2]

Michael Galler's death would have gone unnoticed by anyone other than his family and friends had it not been for Dr. August Wedekind. Three months after Galler's death, Dr. Wedekind went to New York police carrying a thousand-dollar bill Elizabeth Galler had given him. According to Wedekind, in January 1868 Elizabeth had come to his medical office and had explained that she was unhappy in her marriage and wanted to have her husband killed. She wanted the doctor to sell her a poison that would kill her husband. Wedekind later testified that when he "indignantly refused" this request, Mrs. Galler asked him to keep their conversation private and promptly left his office. Wedekind did not hear of Mr. and Mrs. Galler again until November, when he visited another apartment in the building in which the Gallers lived. At that time Wedekind heard from a neighbor that Michael Galler had died a mysterious death from excessive vomiting.[3]

In his version of the events that followed, Wedekind claimed that when he learned of Michael Galler's mysterious death he became suspicious and wrote a note to Mrs. Galler reminding her of their (alleged) conversation in January. Mrs. Galler responded promptly and arrived at Wedekind's office on Orchard Street the very same evening. She came alone and offered to "pay well" if the doctor would keep quiet about all that he knew. When Wedekind said he would not hide his suspicions "for thousands of dollars," Mrs. Galler said she would pay him a thousand dollars in hush money. Two days later, on the evening of Friday, November 12, Mrs. Galler and her brother-in-law came to Wedekind's office bearing a thousand-dollar bill. She gave Wedekind the money and asked for a signed receipt in return. The next morning Wedekind went to Galler's apartment, apparently to inform her that he was going to the coroner's office with his suspicions and the thousand-dollar bill.[4]

A short time after Wedekind told officials his story, New York police had Michael Galler's body exhumed. R. Ogden Doremus, a prominent New York chemist, then examined the corpse for the residue of poisonous compounds commonly used in homicides. Doremus tested specifically for organic poisons such as strychnine and inorganic poisons such as arsenic and corrosive sublimate, but he found none. The only other agents found in Galler's body were morphine and lead. Trace amounts of morphine, which Galler had ingested on the advice of Dr. Dupuis, were found in the victim's stomach and intestines. The lead, however, was dispersed throughout Galler's body. It was found in a black substance that lined the stomach; it was found in the small intestines; it was found in the muscle tissue; and it was found in the liver.[5]

Was it possible that Elizabeth Galler had poisoned her husband using lead? While rare, it was not unheard of to use lead as a means of poisoning. For example, there were cases in which women tried to murder family members by spiking drinking water and food with lead.[6] But Doremus believed that the amount of lead in Galler's body was inconsistent with deliberate homicide. Based on the data reported in newspapers at the time, Galler's corpse contained less than a grain of lead. Autopsies of lead-poisoning victims typically revealed lead levels 3–10 times higher. Doremus also argued that while alive, Galler exhibited none of the more common symptoms of lead poisoning, such as paralysis, a blue gum line, and colic. Furthermore, if someone had tried to poison Galler, the murderer probably would have given the victim large quantities of lead over a short time period, and this would have caused all of Galler's fecal matter to become black with sulfide of lead. Because Doremus failed to observe such a result in his examination, he reasoned that the lead Galler absorbed must have been administered gradually, in small amounts spread over a long period of time.[7]

Following the autopsy, New York officials declared that Michael Galler "came to death from causes unknown to them." The coroner then issued a statement "honorably discharging" Mrs. Galler, even though she had never been formally arrested, and declaring that "there was not, in his opinion, the slightest suspicion against her." A few days later, New York police arrested Dr. Wedekind and charged him with extortion.[8]

But what exactly killed Michael Galler? Although there is no clear answer to this question, Ogden Doremus offered the following hypothesis to the reporters who interviewed him after he performed Galler's autopsy. According to the chemist there was enough lead in Galler's body to produce death if, in terms of overall health, "the patient was very low." As for how Galler had ingested the lead, the chemist could only speculate that since much of New York's water was "partaken through lead pipes" this was an "avenue through which" lead might have gained "access to the system."[9]

The possible link between lead water pipes and Michael Galler's death did not generate universal concern. In its reporting on the case, the *New York Times* said nothing about lead water pipes, and the paper's published accounts do not even include the comments by Ogden Doremus linking Galler's lead intake to the city's use of lead pipes.[10] The *New York Herald* did consider the broader implications of Michael Galler's death, however, and raised these issues in an editorial shortly after the autopsy: "On an analysis of the stomach of the deceased... no traces of organic, acid or mineral poison were discovered, but there were sufficient evidences of the poisonous action of lead, which... might be attributed to the use of [city] water." The paper concluded with a plea for the city to study the health effects of lead water pipes and to search for a safer piping material: "But what about the... lead pipes? Enough is now known to cause a scientific investigation to be made as regards to dangers arising from their use, and what other healthier... pipes can be substituted in their place. This is an important question, and ought to enlist the immediate attention of the Board of Health." As will be made clear, the *Herald*'s plea elicited a limited response from the Board of Health.[11]

The idea that lead water pipes contributed to Michael Galler's death raises at least two important questions. First, if Michael Galler had been exposed to unhealthy amounts of lead, why had his doctor failed to observe any of the more common symptoms of lead poisoning? On this score, one is tempted to challenge the competence of Galler's doctor, Marlin Dupuis. Perhaps Dupuis missed something important. By his own admission, Dupuis had never completed medical school, and the autopsy of Galler's corpse revealed no evidence of a bleeding ulcer, which Dupuis had originally identified as the cause of death. There was evidence that Dr. Dupuis was unable to write his own name. And a twenty-first-century observer cannot help but wonder about the so-called medicines administered to Galler: ammonia and sulphuric acid. But by the standards of 1870, Galler's medical treatment was not all that bad; most doctors at this time used chemicals like ammonia and sulphuric acid in an effort to reduce fevers and destroy pathogenic agents.[12]

Regardless of the shortcomings of Galler's medical care, it is possible to attach undue significance to the absence of a blue gum line, paralysis, and colic. Not all, or even most, adult victims of lead poisoning exhibited such symptoms. Even those prominent and wealthy enough to afford the best medical care were often ill for long periods of time before their physicians were able to uncover the true cause of their suffering. Consider the death of R. Milton Speer in 1890, a former Congressman and important member of the Democratic Party. According to accounts in the popular press, Speer "had been suffering for a year with a nervous disease which baffled the skill of physicians." Only toward the end of his life was it discovered that the disease "resulted from lead poisoning due to drinking water which had long stood in lead pipes."[13]

Second, is there evidence to suggest that Michael Galler's tap water contained unduly high lead levels? The published accounts of Galler's autopsy and the finding of significant lead in his system prompted New York authorities to conduct a few suggestive experiments. In the most revealing of these experiments, the chemist for the Metropolitan Board of Health tested the tap water in his own home for lead. He found that it contained 0.11 grains per gallon, or 1.88 parts per million (ppm).[14] To put this in perspective, the modern EPA standard states that tap water should contain no more than 0.015 ppm—the chemist's tap water had lead levels that were 125 times greater than the current standard. Unfortunately, it is difficult to know how representative the chemist's tap water was of other households in New York City, and for reasons known only to the Board of Health, no broader, systematic study of lead levels was undertaken. Perhaps other household taps carried less lead, perhaps more.

The next time published data on lead levels in New York's water supply appeared was in 1936, more than half a century after Michael Galler's death. In this case, two researchers at Long Island University ran several experiments to estimate the amount of lead New York City water would dissolve from the interior of water pipes. From today's perspective, their findings are startling. When New York water was allowed to remain in service pipes for more than a few days, it would have routinely dissolved enough lead so that water from taps contained about 4 ppm, 267 times the EPA standard and 40 times the level recommended by the United States Public Health Service in 1936. In light of these findings, the *New York Times* ran a very short story in which it recommended that homeowners in the city flush their pipes when returning home from summer vacations. The story was printed on page 21.[15]

New York City did not take any steps to reduce lead levels in its water supply until 1992, 123 years after Michael Galler's death. In that year, the city began treating the public water supply with chemicals to help limit the amount of lead leached from the interior of old water pipes.[16] The city's decision to treat its water took place long after the most serious damage had probably already been done. By 1992, there were relatively low lead levels in New York City tap water, at least by historical standards, and there were only a handful of buildings in the city whose levels exceeded federal guidelines. The reasons for this are simple. Over the course of the nineteenth and twentieth centuries, much of the old lead pipe had corroded away and had been replaced with pipe made of polyvinyl chloride (PVC) or iron, while the lead pipe that remained had developed a protective coating on the interior of the pipe. As a result, in 1992 (prior to water treatment) the building in New York City with the highest lead level in its tap water exceeded the modern EPA guideline by only a factor of 3. In contrast, the statistics cited earlier indicated that between 1870 and 1940, lead levels in New York tap water exceeded the modern EPA guideline by a factor between 100 and 200.[17]

Unlikely Patterns

Michael Galler's death was not the first time people associated a mysterious illness in New York City with the use of lead water pipes. Such associations began in 1848, when New York City finished construction of the aqueduct bringing water from the Croton River in Westchester County to the city. At forty-one miles long, the Croton aqueduct was described as a "sublime engineering feat" that promised to bring the city an unending supply of pure water, free from the taint of disease. Prior to the introduction of the Croton water supply, residents had to rely heavily on surface wells scattered throughout the city. Surface wells were often polluted by nearby privies and cesspools, and were therefore an excellent breeding ground for cholera and typhoid.[18] Because water from the Croton River was largely free of such bacteriological pollution, its introduction helped reduce outbreaks of these and other waterborne diseases. The annual death rate from typhoid fever fell from 6.1 deaths per 1,000 persons before the introduction of the Croton water supply to 2.6 immediately afterward, a reduction of more than 50 percent.[19]

But soon after the Croton water was brought to the city, some physicians began "observing anomalous derangements of the system, and obscure neuralgic arthritic and gastritic affections [sic]." The symptoms appeared consistent with "the slow action of a metallic poison [and] could not be referred to any" source other than water.[20] Dr. Chilton, a chemist, recounted an example when he had been called to examine the water taken from leaden pipes in a house in the city. Several people in the house had become "seriously ill," and Chilton found lead in the water. According to Dr. Chilton, "the effect of lead from drinking of Croton water under such circumstances, is of frequent occurrence, but not recognized as such by the physicians, or rather not attributed by them to the true cause."[21] Although Chilton was a chemist and not a doctor, subsequent observers also claimed that physicians were underdiagnosing the frequency of water-related lead poisoning in the city.

In 1851, George H. Kingsbury, a New York City physician, published a short article in *The New York Journal of Medicine*. Kingsbury described four cases of lead poisoning he had recently treated. In each of the cases, the doctor traced the source of the poisoning to lead-contaminated tap water. The first case involved a middle-aged physician living in the city who had been suffering from an odd constellation of symptoms, including severe abdominal pain, constipation, jaundice, nausea, diminished appetite, rapid weight loss, sleeplessness, and irritability. The patient had visited prominent physicians throughout the city in search of a diagnosis and cure. One thought he had cholera, another thought "biliousness" (liver problems), and another suggested the patient was a hypochondriac. Eventually, one doctor discovered a blue gum line in the patient's mouth, a telltale sign of lead poisoning, and suggested the patient discontinue his use of city tap water. The patient's condition quickly improved after he stopped drinking city water, but returned when the patient, who had not fully believed the diagnosis, began drinking tap water again.[22]

Kingsbury's article failed to convince the broader medical community in New York City. Two years after Kingsbury published his article, the Academy of Medicine met in New York City. An entire session was devoted to a discussion about lead levels in New York City tap water. The session was opened with the comments of one Dr. Joseph M. Smith. After reviewing lead's many effects on the human system and various exposure vectors, Dr. Smith asserted that New York's water was perfectly safe and free of harmful levels of lead. Most of the other doctors at the conference shared Smith's view that Kingsbury was mistaken and that there were no cases of water-related lead poisoning in New York City.[23]

The response of the New York Academy of Medicine, however, was tame in comparison to the rebuke Kingsbury received from Dr. Meredith Reese, the editor of the *New York Medical Gazette and Journal of Health*. Sarcastically describing Doctors Kingsbury and Chilton as "medical savants," Reese claimed his colleagues suffered "under a monomania on the subject of lead poisoning." Reese then characterized Kingsbury's patients as hedonists whose illnesses stemmed mainly from too much food, drink, and sex. "We have known some of them," Reese wrote, "to ignore the effects of high living, generous wines, and still more mischievous excess in sensual indulgence as sources of disease." While not all of Kingsbury's patients took hedonism to extremes, those who did not were "noted hypochondriacs."[24]

One summer night in 1861, twenty prisoners confined in the Kings County Jail in Brooklyn began vomiting uncontrollably. Over the next few days, another thirty or so prisoners developed the same fits of vomiting. The jailhouse physician, Dr. Charles Van Zandt, was summoned and after a few hours of puzzlement came to the conclusion that the jail's water, which was transported via a long lead pipe, had become impregnated with the metal and was poisoning the prisoners. He ordered the jail supervisors to find a new water supply immediately and to replace the lead piping. The inmates quickly recovered once they stopped drinking the tap water. A few days later, a worker at the jail who could not believe that the prisoners had been poisoned merely by drinking from the jail taps, "drank plentifully" of the water in a loud "spirit of bravado." In a few hours, he too fell violently ill and began vomiting uncontrollably.[25] Although no one in New York in 1870 seems to have remembered this incident, the sickness that had afflicted the prisoners in 1861 bore a striking similarity to the one that attacked Michael Galler eight years later.

1

The Significance of the Small

In certain situations, small things can have large effects. For example, in wealthy places with temperate climates, the mosquito is usually little more than a petty annoyance. But put that same small insect in an underdeveloped country with a tropical climate, and the situation is very different.[1] The mosquito proliferates, spreading malaria, yellow fever, and other diseases to epidemic levels. One can tell a similar story about perturbations in climate. While a small change in temperature or rainfall in an open and well-developed economy might do no more than increase food prices by a few percentage points, the same change in an autarkic and subsistence economy might induce famine.[2]

The significance of the small is not limited to the natural realm. One can also see it in seemingly tiny human decisions. This book is a history of one of those decisions. As such, it is also a history of context and contingency, a history of how and why the effects of human choice can vary from place to place. Like the effects of the mosquito, the effects of human decisions are shaped by their larger contexts. There is, however, one important difference between small natural phenomena and the small decisions made by humans. When humans make decisions that prove to be mistakes, we want to understand the source of those mistakes. The desire to understand is even stronger when the same mistake is made repeatedly, has serious public health consequences, and persists despite the consequences.

The Popular and the Arcane

Roughly 150 years ago, cities all over the world installed lead pipes to distribute water. At the time, only a few people gave the decision much thought and, even now, the topic might seem arcane and unworthy of much concern. Yet the decision to install lead water pipes was a fateful one with lasting implications for millions of people. Just as preservatives and chemical pesticides came to be integral and pervasive components of world food supplies, lead pipes came to be integral and pervasive components of urban water systems throughout the world.

A few observations highlight the popularity of lead water pipes. Table 1.1 lists the fifty largest cities in the United States in 1900, and indicates the type of material used for water service pipe in each city in 1897. Service pipes are the pipes that connect homes and apartment buildings to street mains, and are the primary source of metal in household tap water. Of the forty-six cities for which there are data on piping material, all but seven used lead pipe, suggesting that 85 percent of all large American cities used lead in their water distribution systems. Of the twenty-five largest cities, all but two (Baltimore and Kansas City) used lead. In smaller cities and towns, lead piping was less common, but certainly not rare. For example, in U.S. towns with populations less than 8,000 people, one-third used lead pipes, while in towns with populations between 8,000 and 30,000 people, slightly more than half used lead.[3] Unfortunately, there are no accurate data to indicate the frequency with which private homeowners in rural areas used lead to connect their household plumbing systems to private water wells and springs. Anecdotal evidence, however, suggests lead pipes were common in rural areas, and that they were sometimes as long as three-quarters of a mile.[4]

Lead water pipes were not a uniquely American phenomenon. Table 1.2 lists several large cities and county boroughs in Lancashire, England, along with information regarding population and water pipes as of 1890. Lead service pipes were used throughout Lancashire, including the cities of Manchester and Liverpool. Furthermore, in the towns of Bury and Rochdale it was common to find households using lead cisterns to store water. If one takes a broader perspective, other large cities in Great Britain that used lead service pipes include Aberdeen, Dublin, Edinburgh, Glasgow, London, and Sheffield. If one takes a still broader perspective, large cities in continental Europe that used lead pipes include (but were not limited to) Amsterdam, Berlin, Brussels, Haarlem, Leipzig, Lisbon, Madrid, and Paris.[5]

Table 1.1
Lead pipes in the fifty largest American cities, 1900

City	Population	Pipes
New York, N.Y.	3,437,202	Lead
Chicago, Ill.	1,698,575	Lead
Philadelphia, Penn.	1,293,697	Lead*
St. Louis, Mo.	575,238	Lead
Boston, Mass.	560,832	Lead
Baltimore, Md.	508,957	Iron
Cleveland, Ohio	381,768	Lead
Buffalo, N.Y.	352,387	Lead
San Francisco, Calif.	342,782	Lead*
Cincinnati, Ohio	325,902	Lead
Pittsburgh, Penn.	321,616	Lead
New Orleans, La.	287,104	Lead
Detroit, Mich.	285,704	Lead
Milwaukee, Wis.	285,315	Lead
Washington, D.C.	278,718	Lead*
Newark, N.J.	246,080	Lead
Jersey City, N.J.	206,433	?
Louisville, Kent.	204,731	Lead
Minneapolis, Minn.	202,718	Lead
Providence, R.I.	175,597	Lead
Indianapolis, Ind.	169,164	Lead
Kansas City, Mo.	163,752	Iron
St. Paul, Minn.	163,065	Lead
Rochester, N.Y.	162,608	Lead
Denver, Col.	133,859	Lead
Toledo, Ohio	131,822	Lead
Allegheny, Penn.	129,896	Lead
Columbus, Ohio	125,560	Lead*
Worcester, Mass.	118,421	Cement
Syracuse, N.Y.	108,374	Lead
New Haven, Conn.	108,027	Iron
Paterson, N.J.	105,171	?
Fall River, Mass.	104,863	Lead
St. Joseph, Mo.	102,979	?
Omaha, Neb.	102,555	Lead
Los Angeles, Calif.	102,479	Lead*
Memphis, Tenn.	102,320	Lead*

Table 1.1
(continued)

City	Population	Pipes
Scranton, Penn.	102,026	Lead*
Lowell, Mass.	94,969	Lead
Albany, N.Y.	94,151	Lead
Cambridge, Mass.	91,886	Iron
Portland, Ore.	90,426	?
Atlanta, Ga.	89,872	Lead
Grand Rapids, Mich.	87,565	Lead
Dayton, Ohio	85,333	Lead
Richmond, Va.	85,050	Lead
Nashville, Tenn.	80,865	Lead
Seattle, Wash.	80,671	Iron
Hartford, Conn.	79,850	Iron
Reading, Penn.	78,961	Lead

Source: Baker (1897).
* Indicates the city used both lead and iron service pipes.

Table 1.2
Lead pipes in English counties and boroughs, 1888–1889

District	Population	Lead water pipes	Lead cisterns
Barrow-in-Furness	51,712	Limited	No
Blackburn County	119,564	No	No
Farnworth	23,758	Yes	No
Hindley	18,973	Yes	No
Bolton	27,136	Yes	No
Rurnley County	87,058	Yes	No
Bury County	57,206	Yes	Yes
Liverpool City	517,951	Yes	A few
Bootle County	49,217	Yes	A few
Walton	40,304	Yes	A few
Manchester	505,343	Yes	No
Salford County	198,136	No	No
Oldham County	131,463	Yes	A few
Preston County	107,573	Yes	No
Rochdale County	68,458	Yes	Yes

Source: Local Government Board (1888–1889), pp. 379–452.

The health effects of lead water pipes varied from place to place. In many large cities, lead pipes appear to have been used without any serious health effects. For example, evidence presented later in the book indicates that the use of lead pipes in Chicago, London, and Paris was not associated with excess amounts of lead in tap water. However in other cities, using lead pipes appears to have been a serious mistake. Consider recent archeological evidence unearthed in Cape Town, South Africa. During the early 1990s, a team of scientists exhumed the bodies of twenty-eight people who had lived in Cape Town between 1812 and 1922. The scientists examined the teeth of these individuals and found that they contained extraordinarily high concentrations of lead. The tooth-lead levels of many Cape Town residents appear similar to the lead levels observed in modern and well-documented cases of lead poisoning with fairly severe symptoms.[6]

Scientists attributed these high levels of tooth lead to the city's use of lead cisterns and piping to store and transport water. A study of Cape Town tap water conducted in 1914 found that running water had lead concentrations between 0.05 and 0.4 parts per million (ppm), 3–27 times the modern EPA standard. Water that was allowed to stand in pipes or cisterns overnight had lead concentrations between 2.3 and 7.63 ppm, 153–509 times the modern EPA standard. While lead enjoyed widespread use in Cape Town's water system as early as 1812, it was not until the late 1920s that health officials began to discover cases of lead poisoning that they were willing to attribute to the use of city water. Given that the concentrations of lead found in the teeth of Cape Town residents suggest pervasive lead poisoning in the city throughout the nineteenth and early twentieth centuries, it is quite surprising that so few people appear to have been diagnosed with water-related lead poisoning prior to the 1920s.[7]

Cape Town's history prompts many questions. Was the city's water-lead problem an aberration? If not, how common was the problem in other cities around world? On the one hand, the popularity of lead water pipes (see tables 1.1 and 1.2) suggests that most city governments did not consider water-related exposure to lead a serious concern. On the other hand, recent events in Washington, D.C., and Glasgow, Scotland, suggest that water-related lead poisoning was probably not limited to Cape

Town (see chapters 8 and 9). Is it possible that in some cities, water-lead levels were even higher than in Cape Town? Conversely, what explains the contrast between Cape Town and cities like Chicago, London, and Paris where the historical record suggests that lead pipes were used safely? Engineers have long emphasized the importance of various chemical processes in determining the propensity of any given water source to dissolve lead (see chapter 6).

One wonders too about the health effects of regularly consuming tap water with lead levels 500 times greater than the modern EPA standard. What was this doing to the morbidity of Cape Town residents, particularly very young children who were the most vulnerable to lead? Can one adduce any systematic evidence that lead-contaminated water supplies adversely affected health? In an era before widespread screening for elevated blood-lead levels among children, or even among occupationally exposed adults, it is difficult to find such evidence.

Looming behind all of the questions about the health effects of lead water pipes is a more fundamental question. Even water like Cape Town's, which contained lead levels 3–500 times greater than the modern EPA standard, still carried a tiny amount of lead in absolute terms, an amount imperceptible to the naked human eye.[8] How could such a small amount of lead have had any impact on human health? The assumption that the EPA standard is reliable, and that water with lead levels above that standard is unsafe, presumes that the EPA standard is itself an accurate indicator of safety. But what is the scientific basis for the EPA standard? Is it possible that the EPA standard is unduly strict? If not, what sorts of scientific evidence exist to support the claim that even very small amounts of lead might pose a threat to human health? These basic questions mandate a review of the relevant literatures in biochemistry and toxicology.

There are also many questions surrounding the how and why of lead water pipes. What prompted city officials in Cape Town, and most other large cities, to use lead water pipes in the first place? Were alternative piping materials such as iron or cement-line pipe that much more expensive or otherwise unattractive? Were public officials unaware of the dangers of lead? Was the scientific and medical literature on lead so underdeveloped that one could not predict that lead water pipes could pose a serious danger? The difficulty with claiming that people did not know is that the dangers of lead in general, and lead water pipes in particular, were already recognized in the ancient world.[9] But if public officials were at least somewhat aware of the potential dangers of lead pipes, what were the countervailing forces that overrode public health considerations? Recent historical research implicates industry and business interests for their role in marketing lead paint, and in adding lead to gasoline to prevent engine knock.[10] Is there evidence that crude economic interests corrupted the public decision to install lead water pipes? If such evidence is wanting, what other factors might have trumped public health concerns?

Perhaps the most puzzling aspect of Cape Town's experience is that it appears to have taken more than a hundred years for city officials to recognize that the city's water supply contained unhealthy lead levels. Why did it take so long to discover the problem? One might argue that the health effects of excess water lead were trivial, and that the costs of ignoring the problem were small. But nineteenth-century physicians presented evidence that water-lead levels like those observed in Cape Town often resulted in serious and sometimes life-threatening conditions (see chapter 5). Another possibility is that Cape Town's slow response was not representative of most places. But there are other cities where public officials also waited more than one hundred years before responding, in any way, to the unhealthy lead levels found in municipal tap water (see chapter 8).

The Great Lead Water Pipe Disaster offers answers to these questions, and it does so within the context of a broader history of lead water pipes in the modern world. The central arguments of this history are threefold. First, in certain regions, lead water pipes had serious public health consequences, particularly for very young children, the unborn, and childbearing women. For example, the available evidence indicates that in Massachusetts and the north of England lead water pipes increased infant mortality rates and stillbirth rates by between 8 and 25 percent. Another way to illustrate this argument is to compare the lead levels in tap water to the lead levels found in black-market abortifacients. During the late nineteenth and early twentieth centuries, women of child-bearing age sometimes purchased pills made of lead plaster to induce abortion and/or disrupt menstruation. In several towns in Massachusetts one need have consumed only 10–20 ounces of tap water per day to have ingested the same amount of lead as was contained in the recommended daily dose of these abortion pills. Although one cannot be certain of the efficacy of these pills, there is evidence from nineteenth-century medical journals that leaden abortion pills not only contained enough lead to end fetal life, but possessed sufficient lead, in at least some cases, to harm the mother. That pregnant women in Massachusetts and the north of England routinely and unknowingly consumed similar amounts of lead through their tap water corroborates this line of thought (see chapter 3).

Second, the adverse health effects of lead water pipes varied. Water-lead levels in parts of New England and Yorkshire often exceeded the modern EPA standard by a factor from 100 to 1,000, while in the American Midwest and the English South and Midlands, water-lead levels were usually well below the thresholds considered safe by historical observers. The interregional variation in water-lead levels was driven mainly by differences in the chemical characteristics of water supplies across regions. In some regions, water supplies were more acidic and corrosive than supplies in other regions. The more corrosive the water supply, the more the lead leached from the interior of water pipes. The extent to which a water supply dissolved lead was determined by an array of environmental variables, including atmospheric pollution levels, a water supply's chemical and organic content, water treatment techniques, water temperature, and the broader geophysical context (see chapter 6).

Third, the decision by public authorities to install lead pipes, and to continue to use them despite serious public health consequences, resulted from a complex interplay of social forces and scientific conventions. For human societies to adapt and respond to environmental problems in constructive ways, they first need to be aware that there is a problem. The difficulty for those concerned with lead contaminated water supplies is that historically such awareness was limited (though certainly not entirely absent). As explained in later chapters, officials in the nineteenth and early twentieth centuries used the health of adults and older children to assess the effects and dangers of lead water pipes. Simply put, if many adults and older children were being made manifestly ill by lead pipes, the pipes were considered unsafe; if not, officials concluded that the pipes were safe. The difficulty with this approach was that adult health was a poor barometer of water-lead levels. To the extent that water-related lead exposure manifested itself in overt physical symptoms in older populations, it usually did so subtly and slowly, making it all but impossible for even the most acute observers to detect the problem (see chapters 2 and 5).

The prevalence of infectious diseases during the nineteenth and early twentieth centuries further complicated the situation. Infectious diseases, which killed and sickened people quickly and in easily identifiable ways, often blinded observers to the problem of lead-contaminated water, which undermined adult health more slowly and subtly. What was a little premature rheumatism or colic among older persons (common symptoms of prolonged and low-grade lead exposure) when juxtaposed with a disease like cholera, which could kill a person in forty-eight hours and often erupted in epidemics that killed thousands of people within a few months' time? Although the answer to this question might seem obvious at first glance, there is more here than meets the eye. If water lead was inducing rheumatism and colic among adults, it was likely having more serious effects on the very young and the unborn.

Historical modes of understanding also play a central role in understanding why cities used lead. Particularly important is a nineteenth-century engineering principle known as the *doctrine of protective power*. According to the doctrine of protective power, lead pipes could be used safely when the associated water supply was hard or otherwise encouraged the formation of an impermeable coating on the interior of lead pipes. Although the broad contours of this doctrine were correct, it failed to recognize an array of intervening variables other than water hardness which might have also influenced the lead solvency of any given water supply. Furthermore, the doctrine was applied asymmetrically. Cities with water supplies that the doctrine identified as unsafe simply ignored it, or invented their own ostensibly scientific justifications for using lead; those cities with water supplies the doctrine deemed safe applied it blindly, without considering the underlying complexity and randomness of the problem (see chapter 6).

The use and misuse of the doctrine of protective power was driven, in part, by a strong practical incentive. Health effects aside, lead pipes were the most durable and longest lasting piping material available during the nineteenth and early twentieth centuries. Moreover, over the course of the nineteenth century, the doctrine of protective power came to be interwoven with a political movement known as municipal socialism. When science and political ideology were combined it changed the way many engineers and policymakers viewed the doctrine of protective power; they came to hold the doctrine more as a matter of religious conviction than as a scientific hypothesis subject to empirical verification and revision. The repercussions for water consumers were not inconsequential. Often they too came to believe that water that passed through lead pipes could never absorb sufficient amounts of lead to become poisonous (see chapter 6).

Finally, the interaction between market institutions and the law was pivotal in determining the incentive structure surrounding lead water pipes. During the nineteenth and early twentieth centuries, the courts held water consumers, as opposed to suppliers, liable for all damages resulting from lead-contaminated tap water. This created incentives for consumers to prevent lead-related diseases through private means, but it undercut incentives for the operators of large public water systems to do the same. In part because water suppliers could have more easily and cheaply invested in the technologies necessary to prevent water-related lead poisoning, the legal system encouraged costly and relatively ineffective modes of disease prevention. The ill effects of this incentive structure were compounded by the fact that consumers typically had poor information regarding the safety of lead pipes. Furthermore, even for the small subset of consumers who were aware of the dangers of lead, local ordinances often prevented them from adopting safer piping materials like iron and cement (see chapter 7).

Evidence and Approach

The Great Lead Water Pipe Disaster draws heavily from traditional historical sources, including government reports and investigations, important legal cases, articles from nineteenth-century medical journals, books from the same era, and accounts from major newspapers. Although there is ample evidence in these sources alone that lead water pipes constituted a serious public health problem during the nineteenth and early twentieth centuries, these sources will be supplemented with a review of the most recent scientific research on lead poisoning and with the application of modern statistical techniques to historical data. Supplementing traditional forms of historical evidence with modern methods of analysis allows for the construction of a more rigorous and scientifically sound assessment of water-related lead poisoning in history. It also enables the evaluation of the reliability of the qualitative evidence.

As a means of illustrating the power of this approach, consider the following turn-of-the-century articles and books addressed in subsequent chapters. Chapter 3 discusses an article by a British physician named Alfred Swann. In an article published in the *Lancet* in 1892, Swann claimed that water-related lead poisoning in England represented a serious danger to the country's health and welfare. To bolster his case, Swann cited the case histories of several of his patients who had been poisoned by lead-contaminated tap water. The problem with citing Swann's article alone is that it is difficult for the reader to judge its generality and scientific plausibility. Was Swann overstating his case? Were lead pipes really a national danger, or merely the cause of an isolated illness here and there? The case histories presented by Swann cannot answer these questions. However, by comparing infant mortality rates and other health outcomes across cities with high and low water-lead levels one can quantify the size of the problem. Once the quantification is done, readers can judge for themselves if Swann was exaggerating.

Similarly, much of chapter 4 is dedicated to discussing a book published by Norman Porritt, a prominent British surgeon during the late nineteenth and early twentieth centuries. Porritt argued that the excessive rates of pregnancy-related seizures and comas in certain parts of Britain were correlated with inflated lead levels in local water supplies. Porritt also presented data which showed that in those parts of England where water-lead levels were high, so too were the rates of pregnancy-related comas and seizures, while in those areas with low water-lead levels the incidence of comas and seizures during pregnancy was also low. Although Porritt's analysis was state-of-the-art for its day, modern readers will question two aspects of his analysis. First, what is one to make of Porritt's argument that *prior* lead exposure increases a woman's risk of coma or seizure during pregnancy? Is there any modern scientific evidence to suggest such a possibility? Second, how trustworthy was Porritt's crude ecological analysis? Because Porritt was unable to control for potentially confounding variables, it seems possible that the correlations he identified were spurious. A review of the relevant medical literature can answer the first set of questions; a formal statistical analysis can address the second.

The supplementary evidence—that is, the modern scientific research and the statistical analyses—will be presented in two different formats. Given the importance of the modern scientific literature on lead poisoning for this book's larger arguments, a review of this literature is presented in chapter 2. Because the statistical work will be less accessible to many readers, and because it might detract from the continuity of the narrative, it will be presented in appendices. Only key statistical results will be discussed in the body of the text. Readers wanting to know the details of the estimation strategies and data sources can turn to the appendices. In general, the statistical results are predicated on techniques commonly used by health economists. The data employed are mainly from late-nineteenth- and early-twentieth-century Massachusetts and Great Britain, two regions known for having had reliable birth and death registration systems.[11]

Lead Water Pipes as an Environmental Problem

Ask people today to list the greatest environmental disasters of the last two hundred years and most would focus on recent events such as Chernobyl, Bhopal, the destruction of the rainforest, or Love Canal. More historically minded individuals might also mention the Industrial Revolution, the diversion and destruction of natural lakes and rivers to provide water to large cities like Los Angeles and San Francisco, or the introduction of the automobile. Whatever the merits of these claims, almost no one would mention the introduction of lead water pipes as a significant event in environmental history. Yet lead water pipes killed or harmed many more people than were injured by events in Bhopal, India, or at Love Canal. As mentioned previously, cities the world over used lead pipes to distribute drinking water to residents. In this way, lead water pipes were like the Industrial Revolution or the automobile: they affected the lives of millions of people around the world. Perhaps as many as eight million people were affected by an epidemic of water plumbism (lead poisoning, from L. *plumbum*, lead) in the North of England during the 1880s.

Although lead-contaminated water, narrowly construed, is a much smaller public health problem today than it was a hundred years ago, the history of water lead has implications for a broader set of concerns surrounding drinking water and public health.[12] For example, there is evidence that exposure to toxic metals through drinking water remains a worldwide problem, and that such exposure can have serious public health consequences. Probably the best-known case comes from twentieth-century Bangladesh, where thirty-five to fifty-seven million people (27–44 percent of the country's population) had been consuming arsenic-contaminated water for decades without being aware of it. The most serious victims suffer from skin lesions, gangrene, and various cancers.[13] Furthermore, there are concerns that high arsenic levels in water supplies in China, Finland, the United States, and Taiwan are correlated with excessive cancer rates in some regions.[14] In addition to arsenic, metals such as boron, cadmium, chromium, cobalt, copper, nickel, and selenium have been identified as potentially significant pollutants of modern water systems.[15]

There are strong parallels between the history of water and lead and what has happened with these other toxic metals. For example, there is evidence that Bangladeshis are poorly informed about the dangers of arsenic in drinking water and that they are unaware of the most effective modes of protecting themselves from exposure.[16] There is also evidence that the effects of toxic metals like arsenic, at low levels of exposure, take years to develop and are difficult to isolate and measure. As with lead, however, diagnostic problems do not mean that the toxic effects are not present.[17] Furthermore, the regulatory framework governing exposure to heavy metals in drinking water is incomplete in many parts of the world, meaning that consumers may be regularly exposed to dangerous amounts of such metals.[18] It is also possible that as more is learned about these

metals, policymakers will periodically reduce what is considered an acceptable level of exposure—as was the case with lead in drinking water.

Lead Water Pipes and the Mortality Transition

During the early twentieth century, the United States saw rapid improvements in life expectancy, and improvement was especially pronounced in
the heavily urbanized Northeast. Reductions in infant mortality account
for most of the improvement in life expectancy.[19] What caused the transition from high to low mortality, in the United States and elsewhere, has
been the subject of extensive academic debate among historical economists and demographers. Typically, scholars emphasize one or more of
three forces as the engine behind falling infant mortality. For some, it
was the diffusion of the germ theory of disease to the general population.
Increased knowledge of appropriate sanitary practices enabled parents to
adopt a variety of behaviors that helped protect infants, including breast-
feeding and boiling tap water.[20] For others, it was investments in infrastructure related to public health, such as water and sewer systems, that
helped protect infants.[21] For still others, it was economic growth and the
associated improvements in nutrition that explain the urban mortality
transition.[22]

The results presented here suggest that historical demographers need
to think about lead eradication, particularly as it relates to public water
systems, as a source of improved infant mortality rates. The results here
also highlight the need to think about region-specific causes of demographic change. Because water supplies in Massachusetts and northern
England were corrosive relative to supplies elsewhere in the United States
and the United Kingdom, these areas were affected more severely by the
practice of using lead pipes to distribute water.

Lead and Lead Poisoning in History

There already exists a large literature on the history of lead and lead
poisoning. The extant literature has two strains. One strain is predominantly archeological and scientific, and seeks to identify long-term
changes in human lead exposure from the ancient world to the present;

the other strain uses techniques from social and medical history, and focuses mainly on the past two or three centuries.[23] Straddling these two strains of the lead literature, this book makes two contributions. First, it brings new analytical methods to bear, particularly the use of modern statistical techniques. These techniques make it possible to estimate the magnitude and public health effects of lead poisoning in human history. Although some of the results here have already been published in environmental science journals and other outlets,[24] this book brings the results together in a systematic way and makes them accessible to a broad range of readers.

Second, this book focuses on a source of environmental lead exposure that has received little attention, and it focuses on that source over a long period of time. This orientation reveals new insights into the medical and social history of lead poisoning. Some of these new insights relate to the work of medical researchers like Alfred Swann and Norman Porritt, who were among the first scientists to draw attention to the correlation between lead and reproductive health. Other insights relate to political economy. In particular, recent histories of lead emphasize the influence of industrial interests in shaping the regulatory policies governing leaded gasoline and lead paint; those interests play a much smaller role in the history developed here. The orientation of this book is toward exploring how and why one source of environmental lead exposure has changed over time. The central finding is that in the era before leaded gasoline and the widespread use of lead paints, water was likely the primary source of lead exposure for individuals living in regions with corrosive water supplies and lead pipes.

2

A House for Erasmus

Erasmus D. Fenner was a prominent physician in the city of New Orleans. A researcher and a frequent consultant to the local government, Fenner's medical pronouncements usually enjoyed a favorable reception from his colleagues in New Orleans and the American South in general. There was, however, one episode that failed to further Fenner's standing in the medical profession, and might well have hurt it. Between 1849 and 1851, Dr. Fenner studied an epidemic of colic in New Orleans. Although not usually fatal, colic was associated with painful abdominal cramping and often constipation or diarrhea. Fenner came to believe that the city's colic epidemic stemmed from the widespread use of lead water pipes and cisterns. According to Fenner, there were more than forty-seven miles of lead service pipes in the city, serving around five thousand households and institutions, with each pipe averaging about fifty feet. For those without access to the city water supply, the use of lead cisterns to collect and store rainwater was also common, as was the use of lead tubing in soda fountains.[1]

Fenner presented four pieces of evidence to support his theory. First, there were large increases in the incidence of colic during the years of 1838, 1849, and 1850. These years corresponded with large extensions in the city's public water system, and a concurrent expansion in the number of people who would have been exposed to lead in their drinking water. Second, in addition to colic, there were a host of other unexplained, and potentially lead-related, illnesses in the city, including convulsions, paralysis, neuralgia, and rheumatism. According to Fenner, this pattern of lead-related ailments corresponded perfectly with well-documented outbreaks of lead poisoning and colic, such as the

Devonshire colic in England (which was caused by apple cider made with lead apple presses) and the "dry-bellyache of the West Indies" (which was traced back to lead-contaminated spirits). Third, Fenner presented case-studies of colic-stricken individuals who improved once they stopped drinking city water or lead-contaminated soda water. One particularly unfortunate incident involved a large family on Magazine Street. First reported by another physician in New Orleans, four children in the home died from convulsions, and a fifth developed convulsions and paralysis of the leg. This family used a lead cistern to collect rain-water and a large amount of lead was found in the water from the cistern.[2]

The least compelling evidence presented by Fenner was also his most important. Fenner tried to measure the lead levels in New Orleans tap water. Unfortunately, Fenner appears to have not been a particularly talented chemist. All he claimed was that the water samples became discolored when exposed to sulphuretted hydrogen gas, indicating that the water did contain some lead, and perhaps, though not necessarily, a large amount.[3] This metric, however, only offered a rough qualitative measure of the water-lead level; it did not indicate a precise quantitative level. Fenner also made the mistake of testing only a few samples of water. Summarizing his test results, Fenner wrote: "My experiments were on a small scale, and I may have been deceived, but I saw enough to greatly strengthen my suspicions of the presence of lead" in city tap water.[4]

Whatever the shortcomings of his evidence, Fenner's research did not go unnoticed. After reading one press account, a California doctor wrote Fenner to say that he had discovered a subtle case of lead poisoning as a result of Fenner's research. Around the same time Fenner published his findings, a physician in Lowndsborough, Alabama, published an account of epidemic colic in the town that he traced to lead water pipes. And according to Fenner, two independent doctors in Texas discovered cases of water-related lead poisoning, which prompted the abandonment of lead pipes and cisterns in Galveston.[5] Last, the finding that soda fountains in New Orleans had high lead levels resulted in several establishments' discontinuing the use of lead tubes and posting signs over their fountains that read, "No Lead Pipe." There is evidence that the city council eventually passed legislation prohibiting the transmission of soda water through leaden pipes as a result of Fenner's research.[6]

But these were the accolades of the trivial. Among the groups that mattered and on the questions Fenner considered most significant, the responses were much less enthusiastic. Despite Fenner's urging, city officials did little to study the effects of lead service pipes, or to pass legislation prohibiting them. New Orleans continued using lead service pipes until at least 1897, and likely well into the twentieth century.[7] When Fenner presented his results to the Physico-Medical Society, an association of physicians from the South, "a committee was appointed to investigate the subject" of lead-contaminated drinking water in New Orleans, and to witness "a repetition" of the experiments conducted by Fenner. Made up of one physician and two chemists, the committee rejected Fenner's conclusions on all counts except those regarding lead-contaminated soda water.[8] On the dangers of lead water pipes, the issue that Fenner believed was "by far the most important part" of his inquiry, the committee could only say that it hoped others in the medical society would further investigate the situation.[9]

The rejection of Fenner's conclusions regarding lead water pipes was based on three things. First, if it were the lead water pipes and lead cisterns that were making people sick, more people should have been made ill with colic, because almost the entire population of the city was exposed, one way or another, to lead from these sources. Yet only a fraction of the city suffered from the epidemic colic. In response, Fenner could only say that lead was an idiosyncratic poison, affecting some people much more than others. He cited data from occupational settings showing that among workers all facing a similar level of lead exposure, only a subset of the workers became poisoned. Second, critics wondered about the seasonal variation in colic. How come, they asked, the colic always peaked during the summer months? Fenner responded that people drank more water in the summer, and that in industrial settings, cases of lead poisoning also seemed to peak during the summer.[10]

For Fenner's critics, the most troubling aspect of his research was his chemical analysis of New Orleans water, and his related claim that the water was lead solvent. As stated previously, Fenner was unable to derive a precise quantitative estimate of the amount of lead contained in city tap water.[11] Looking back, it seems unlikely that Fenner would have been more successful in convincing his peers even if he had constructed more precise estimates of water-lead levels. Most scientists and physicians during the mid-nineteenth century held a high threshold for safe lead exposure, and it is not at all clear that the lead levels in New Orleans water would have been proven to be above that threshold.[12] Nevertheless, the rhetorical effect of Fenner's inability to estimate water-lead levels was compounded by his unorthodox understanding of the chemical characteristics that made a particular water supply lead solvent. Fenner claimed that water from the Mississippi River, from which New Orleans drew its supply, contained elements that made the water corrosive. The prevailing wisdom, however, maintained that these same constituents caused an insoluble barrier to form on the interior of water pipes, limiting the amount of lead taken up by the water.[13]

The foregoing discussion suggests that Fenner's colleagues did not reject his arguments because they were corrupt or because they had some bizarre devotion to lead, as will be observed in historical episodes presented later in the book. On the contrary, the available evidence suggests that Fenner's critics rejected his arguments because he could not adequately address their evidentiary concerns. The question at hand is why Dr. Fenner could not address these concerns. Was it just a shortage of evidence and statistics, or was there a more fundamental problem?

The French mathematician Henri Poincaré once said, "science is built of facts the way a house is built of bricks; but an accumulation of facts is no more science than a pile of bricks is a house."[14] When Erasmus Fenner tried to convince his colleagues about the perils of lead, all he had was an accumulation of facts, a pile of bricks without a substructure to support them or mortar to hold them together. Put more formally, without a scientific model of lead poisoning, Fenner could not assemble his evidence in a meaningful and coherent way, nor could he explain the underlying physiological mechanisms that gave rise to the phenomena he observed. As a result, in responding to criticism, or in constructing his own arguments, Fenner could only defend his position by presenting more facts. For example, when asked why water-related lead poisoning had such idiosyncratic effects, Fenner responded by presenting data that showed that occupational lead exposure also had varying effects across the population of lead workers. If Fenner had possessed a scientific model of how lead worked, he would have been able to explain the many physiological parameters affecting an individual's vulnerability to lead.[15] Similarly, Fenner's claim that Mississippi River water was lead solvent would have been strengthened had he been able to explain, in even crude scientific terms, the chemical processes that made water corrosive.

The rhetorical significance of scientific models is also well illustrated by the debate surrounding the seasonal variation in colic. While Fenner's critics maintained that this variation undermined the case that the colic was caused by lead-contaminated water, when viewed in the light of more recent scientific understanding, it probably strengthens the case. Over the past century, chemists and engineers have developed increasingly elaborate models to explain the ability of a particular water supply to dissolve lead.[16] These models show that, holding everything else constant, hot water dissolves significantly more lead than cold water.[17] Also, recent research suggests that blood-lead levels tend to rise during the summer months, and this increase is explained at least partially by variation in environmental exposure to lead over the different seasons.[18] If Fenner could have situated his findings in the context of this research, he would have been able to use the seasonal variation in colic as a supporting piece of evidence, rather than have it be something that needed to be rationalized or explained away.

But to best illustrate the importance of scientific models, consider the following counterfactual scenario. Assume that Fenner had possessed a good set of "facts" for a modern epidemiologist: a controlled study comparing the incidence of colic and other lead-related illnesses among New Orleans residents, who had been exposed to varying levels of lead in their tap water. Assume further that this set of facts showed that the incidence of colic and other lead-related ailments was positively correlated with water-lead levels. In other words, assume that Dr. Fenner had evidence similar to the statistical evidence presented in modern studies of lead poisoning. Without a model of lead poisoning that explains the physiological and biochemical mechanisms that enable lead to induce colic, even friendly observers will still harbor doubts as to the nature of the observed statistical correlation: is the correlation spurious—the result of some confounding variable? Before one accepts the proposition that a particular statistical correlation is causal, there has to be a theory, a body of knowledge, that tells the observer how and why two variables should be correlated. In the case of lead, one needs to know why lead is poisonous and how, exactly, it makes people sick. Fenner could explain none of this.

Accordingly, this chapter reviews the relevant scientific literature on lead poisoning. In doing so, it develops a framework in which to evaluate the evidence presented by historical actors such as Erasmus Fenner. Specifically, the chapter will identify the clinical manifestations of lead poisoning, particularly with regard to reproductive health outcomes, and the molecular targets of lead once in the human body. Two examples from later chapters illustrate the significance of this literature review. First, chapter 3 presents historical evidence that lead-contaminated water supplies adversely affected fetal and neonatal health outcomes. Recent medical research establishes the scientific plausibility of this evidence by showing that even low levels of in utero lead exposure can increase the risk of miscarriage, stillbirth, and neonatal mortality. Chapter 4 argues that it was difficult for nineteenth-century physicians to diagnose *low-grade* lead poisoning in adult patients, and that such patients often had to suffer for many years before receiving an accurate diagnosis. Explaining lead's multisystemic and idiosyncratic effects, this chapter provides a scientific foundation for the argument that low-grade lead poisoning was difficult to diagnose.

The Multisystemic Effects of Lead

Lead affects multiple systems in the human body, including the central and peripheral nervous systems, the gastrointestinal tract, the kidneys, and the hematological system. Which of these systems is affected and to what degree depends in part on how much lead has been ingested and retained by the body. Children are more vulnerable to lead because their systems absorb more lead than do adult bodies, and because their developing systems are less able to withstand the toxic effects of lead. Lead is a cumulative toxin so repeated exposure does not produce immunity; instead repeated exposure, even at low levels, causes the metal to accumulate in the body and produce more severe physiological damage.[19]

Table 2.1
How lead affects children and adults

	Effects	
Blood-lead level	Children	Adults
0–9 μg Pb/dl	Uncertain*	Uncertain*
10–19 μg Pb/dl	Developmental delays; ↘ Vitamin D metabolism; EP[a]	Hypertension; EP[a] (women)
20–29 μg Pb/dl	↘ Nerve conduction velocity	EP[a] (men)
30–39 μg Pb/dl		↗ Systolic blood pressure (men); ↘ Hearing acuity
40–49 μg Pb/dl	↘ Hemoglobin synthesis	Peripheral neuropathies;[b] infertility (men); nephropathy[c]
50–100 μg Pb/dl	Colic; frank anemia; nephropathy;[c] encyphalopathy[d]	↘ Hemoglobin synthesis; ↘ longevity; frank anemia; encephalopathy[d]
>100 μg Pb/dl	Death	Death

Sources: Perazella (1996); Ravin and Ravin (1999); Xintaras (1992); Needleman (2004).
↘ Decreased function.
↗ Increased function.
* See associated discussion in text.
[a] Erythocyte protoporphyrin: changes in the shape and size of red blood cells.
[b] Nerve disorders in the extremities. Historically, such disorders might have manifested themselves as complaints about "rheumatism" in the hands and feet, and wrist- and foot-drop.
[c] Chronic or acute kidney failure.
[d] Any brain-related disorder. Historically, such disorders might have manifested themselves in violent mood swings, memory loss, and dementia.

Table 2.1 summarizes the symptoms implied by varying blood-lead levels for both children and adults. Turning first to the effects on children, at blood lead levels between 10 and 39 μg/dl (micrograms per deciliter), lead is associated with developmental delays, reduced vitamin D metabolism and nerve conduction velocity, and changes in the size and shape of red blood cells. At levels between 40 and 99, lead is associated with more severe pathologies including reduced hemoglobin synthesis, colic, anemia, kidney failure, and brain swelling. At blood-lead levels above 100, the effects of lead are often fatal for children.

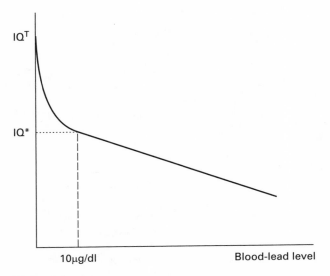

Figure 2.1
IQ and blood lead: Relationships implied by existing research. *Sources:* Lanphear
et al. (2000); Canfield et al. (2003).

Although table 2.1 indicates that the pediatric effects of lead at levels
below 10 μg/dl are unclear, recent research suggests that this is probably
too conservative. More precisely, it had long been thought that as long
as children had blood-lead levels below 10, variation in blood-lead levels
would have no effect on their intelligence. Figure 2.1 illustrates the
implied relationship. The dotted line below the 10 microgram threshold
depicts the assumed ineffectiveness of lead below this threshold; the
declining solid line after 10 depicts the steady erosion in intelligence as
blood-lead levels rise. But recent research by Bruce Lanphear and others
indicates that the marginal or incremental effects of increased lead exposure are actually greatest at the lowest levels of exposure, levels long
believed to have been safe.[20] This new evidence suggests that the relationship between lead exposure and IQ looks more like the solid line
depicted in figure 2.1: a rapid degradation in IQ at blood-lead levels
below 10 and a less rapid, though continuing, decline after this threshold
has been reached.[21]

As for how lead affects adult health, at blood levels less than 20 μg/dl,
lead causes subtle changes in body chemistry and manifests itself in diz-

ziness and hypertension. At blood levels between 20 and 40, lead has more serious effects, including peripheral neuropathies, infertility in men, and increased systolic blood pressure. At high levels of exposure, lead causes nephropathy, frank anemia, and reduced hemoglobin synthesis. At blood levels above 100, lead can cause death. The estimated threshold effects of lead are being reduced continuously; for example, among adults exposed to lead in their occupations, it had long been believed that any exposure level below 50 μg/dl was safe and did not impair cognitive function. Over the past ten years, many observers have started to reconsider this threshold and now recommend lowering it. The basis for this revision is that previous testing procedures did not adequately control for "individual biological diversity" and were not sufficiently sensitive to the subtle and sometimes contradictory effects lead can have on the nervous system. Furthermore, conventional psychological tests are complicated and vulnerable to the confounding effects of human culture. Newer tests, based on recent developments in biochemistry and less vulnerable to cultural variation, indicate that the neurological systems of adults are more sensitive to effects of lead than had been previously considered.[22]

Aside from the pathologies discussed in table 2.1, there is growing evidence that lead exposure is associated with an increased risk of cardiovascular disease. The correlation between lead exposure and cardiovascular disease holds true even for individuals who were exposed to lead as children but not as adults. One study shows that if the mean blood-lead level in the United States could be cut in half, the annual number of myocardial infarctions would fall by about twenty-four thousand, and the incidence of cardiovascular disease would fall by over one hundred thousand. According to the author of the study, this represents "a large attributable risk compared to most environmental toxins." Moreover, the geographic distribution of cardiovascular disease, in both the United States and Europe, is correlated with the lead solvency of public water supplies, although the source of this correlation is subject to multiple interpretations.[23]

Historically, some observers believed that lead exposure lowered the resistance to disease. This belief, however, was predicated largely on simple correlations relating occupational exposure to specific infectious diseases. Because no physiological pathway through which lead might have poisoned the immune system was identified, the hypothesis that lead might function as an immunotoxin never gained wide favor. Consider, for example, the following excerpt from the writings of Alice Hamilton, a physician at Harvard and an industrial reformer who, during the early twentieth century, specialized in lead:

Lead lowers the resistance of the body to infectious agents, especially such infections as tuberculosis and blood poisoning. Certain industries, as, for instance, the typographical trades, have always had a far larger proportion of tuberculosis than can be accounted for in any way except on the ground of a lowered resistance to tuberculosis caused by the absorption of lead. Suppurative inflammations also are more common among lead workers than among men not exposed to lead. The men themselves say that if a lead worker cuts himself the cut always festers, because the lead gets in and poisons the cut. What really occurs is that the germs of suppuration get in and the tissues, being affected by the lead, do not offer much resistance to them.[24]

Without a model to explain how and why lead might have compromised immune function, it is easy to read such passages and speculate that forces other than reduced immunity might have driven the correlation between lead exposure and tuberculosis among typesetters and printers.

Because lead affects so many physiological processes, it can produce a wide variety of symptoms, including vomiting, constipation or diarrhea, colic, flatulence, jaundice, dizziness, hearing difficulty, headaches, fever, epileptic-like convulsions, depression, irritability, anxiety, strong thirst, loss of appetite, anorexia, bad breath, a peculiar taste in the mouth, weariness and lethargy, sleep disorders, vision problems, weakness in the extremities, pain, cramping, burning sensations in the extremities, memory loss, hallucinations, rheumatism, gout, paralysis (especially wrist- and foot-drop), anemia, and menstrual disorders. There are three notable features to this list. The first is the incongruity of some of the symptoms. Some patients develop diarrhea; others are constipated. Some patients are anxious; others lethargic. The second notable feature is the great variety of symptoms, highlighting the fact that lead's effects are multisystemic. The third feature is the generic quality of many of these symptoms. For example, vomiting, diarrhea, lethargy, and anemia are certainly not unique to victims of lead poisoning.

It is important to point out that the symptoms of lead poisoning become more distinctive the more severe the exposure level. At high levels of exposure, lead poisoning induces unmistakable symptoms, such as wrist- and foot-drop and a blue gum line. These symptoms are not usually seen in association with other pathologies, and make it relatively easy to diagnose lead poisoning at high exposure levels. However, at low levels of exposure, the symptoms of lead poisoning are more subtle and generic, such as lethargy, irritability, constipation, hearing loss, and difficulty sleeping. These symptoms are in no sense unique to lead poisoning and are more typically caused by aging, mood disorders, and other sources. As such, low-grade lead poisoning is more easily mistaken for other pathologies than high-grade lead poisoning.

The historical literature on lead poisoning emphasizes the idiosyncratic effects of lead on specific individuals: The same amount of exposure across a given population might induce severe symptoms such as paralysis and insanity among some people while inducing less severe manifestations in others. Sir Thomas Oliver wrote in 1897 that "people are not equally affected" by lead, and that "there is not only an individual idiosyncracy [but also] an hereditary disposition to suffer from lead."[25] Similarly, writing in 1888, W. R. Gowers, a physician at University College, London, maintained that the "occurrence of symptoms of lead poisoning [is] to some extent determined by individual peculiarities" such as nutrition, age, sex, and hereditary factors.[26] Recent research supports the general claim that lead has idiosyncratic effects.

Aside from age and sex, modern research has identified the following factors as determinants of an individual's vulnerability to lead: genes, nutrition, and the broader social environment. In particular, three genes have been identified as possibly shaping an individual's vulnerability to lead poisoning. The first is the genetic coding for the ALAD enzyme, which, as will be explained, is important for the production of hemoglobin. The second is the vitamin D receptor gene, which helps determine calcium absorption in the gut. The third is the HFE gene, which is involved in iron metabolism. In particular, in some individuals the presence of the HFE gene gives rise to hemochromatosis, a disease characterized by liver dysfunction, excess iron in organ tissue, and skin discoloration.[27]

Calcium and iron deficiencies increase the amount of lead that an individual absorbs into his or her body. Animal studies have shown that some dietary "substances bind lead and increase its solubility, thus enhancing its absorption." These substances include sodium citrate (a salt of citric acid); ascorbate (a salt of vitamin C); amino acids, vitamin D, protein, fat, and lactose.[28] Recently, it has also been suggested that antioxidants such as vitamins B6, C, and E, beta-carotene, selenium, and/or zinc might help prevent and/or treat lead poisoning.[29]

Lead and Fetal Development

One of the greatest puzzles confronting physicians well into the twentieth century was the relationship between lead exposure and human reproduction. As will be explained in the following chapter, doctors as early as the 1860s began documenting cases of infertility and fetal/infant death that appeared to have some connection to lead exposure. The difficulty confronting historical observers was that they knew very little about how and why lead influenced fetal health outcomes. Without such knowledge, doctors concerned about undue lead levels in public water supplies had a hard time convincing more skeptical observers that such lead posed a serious health hazard. Consider, for example, the experience of Alfred Swann, who will be discussed in much greater detail in chapter 3. In 1889, Swann presented evidence that lead water pipes were associated with sterility and spontaneous abortion. There were two problems with his evidence. First, it was not based on a systematic study, but on a series of case-studies. His evidence was vulnerable to skepticism about how accurately it reflected a larger sampling. Second, even if Swann had possessed the statistical tools and raw material necessary to conduct a full-blown ecological study, skeptics and critics could argue that his correlations were spurious. What Swann needed was a scientific construct, a larger framework to explain how and why lead had such disastrous effects on human reproduction.

Recent studies indicate that in utero lead exposure is associated with poorer fetal and neonatal health outcomes. One recent study found that lead levels in the bones of stillborn and nonsurviving neonates were 5–10 times higher than the bone-lead levels of normal infants.[30] While this same study found lead levels were higher in malformed infants than in normal ones, most researchers believe there is not sufficient evidence to justify the claim that lead produces developmental malformations.[31] Nevertheless, the basic claim that maternal lead exposure during pregnancy increases the risk of spontaneous abortion, stillbirth, and neonatal death has been corroborated in several other independent studies.[32] There is also evidence that lead exposure is associated with reduced birth weight.[33] To the extent that low birth weight, and poor infant health in general, is associated with an increased risk of cardiovascular disease and other chronic conditions later in life, in utero lead exposure might have lasting health effects even on those infants who survive the initial exposure.[34] Recently, researchers have shown that mothers with a history of childhood lead exposure have higher rates of spontaneous abortion, stillbirths, and living children with learning disabilities.[35]

While the existing literature establishes a link between in utero lead exposure and fetal development, several important questions remain. We need to know more about threshold effects, the precise nature of the dose-response relationship, the mechanisms through which lead poisons the fetus, and the extent (if any) to which lead might function as a teratogen.[36] Of special concern for this discussion is the debate surrounding the existence of a threshold effect. Until the last decade or so, researchers were generally skeptical of the proposition that low levels of lead exposure could induce abortion, although a correlation at higher levels of exposure was accepted.[37] The most recent epidemiological evidence, however, suggests the scientific ambiguity that has surrounded low-level effects is the result of poor data and methodology. Improvements in data and methodology yield results that imply that even low-level lead exposure can induce abortions and stillbirth.[38]

To highlight the importance of experimental design and appropriate statistical techniques in isolating the effects of low-grade lead exposure on fetal development, consider the following: A study of women in Mexico City conducted during the late 1990s found that for every 5 µg/dl increase in blood lead, the risk of a spontaneous abortion rose by 80 percent and this relationship held true even at low levels of exposure. This particular study included 668 subjects and was specifically designed to overcome the problems that had hampered previous work on the subject. In previous studies, researchers had failed to adequately solve the following difficulties: there were not significant differences in the blood-lead levels of the test cases and the controls; no effort was made to control for the large increases in blood lead that occur during pregnancy; there were small sample sizes and low response rates in those studies predicated on surveys; and there were often inadequate controls for confounding variables, such as employment.[39]

Lead and Eclampsia

Eclampsia is coma or seizure activity in pregnant women with no prior history of such activity. The disease almost always strikes during the third trimester of pregnancy or within two days after delivery. It is more common in first-time pregnancies than in later pregnancies. Except in rare cases, eclampsia is preceded by preeclampsia, a condition manifested in hypertension, abnormalities in kidney function, and metabolic changes such as fatigue and increased body temperature. The current state of medical knowledge regarding eclampsia and preeclampsia is incomplete and dominated more by theory than certitude. While there exists a rudimentary understanding of the factors that might predispose some women to eclampsia, there is no cure for the disease, nor is there a precise understanding of what induces eclampsia, or its antecedent, preeclampsia.[40]

Although the causes of eclampsia are probably manifold, undue lead exposure appears to be a contributing factor in at least some cases today. In particular, recent research suggests that eclampsia is related to the level of various metals in the mother's system, and that too much or too little of any of these metals might induce eclampsia, or its less severe antecedent, preeclampsia. During the later stages of pregnancy, metals are mobilized from the mother's skeleton and enter her bloodstream and soft tissues. The mobilized metals include copper, zinc, sodium, potassium, calcium, and magnesium. At appropriate levels, these metals are necessary and desirable for fetal development. Recent studies, for example, suggest that deficiencies in calcium, magnesium, iron, and zinc might be associated with fetal and/or maternal pathologies.[41] Unfortunately, when these metals are mobilized from the skeleton a potential problem arises:

Along with the desirable metals, toxic metals such as lead and cadmium are also mobilized. Because 90 percent of all lead in the human system is stored in the bones, the shock to the system when skeletal lead is mobilized can be significant. In a recent study of one hundred pregnant women in Australia, it was found that mobilization of skeletal lead increased blood-lead levels by an average of 31 percent, and by as much as 65 percent. Even in women with relatively low lead levels, there were large effects. What makes this Australian study particularly important is that it focused on immigrant women, who while in Australia were exposed to little, if any, lead. Nearly all of the lead in their systems was the result of lead exposure prior to migration and pregnancy. This suggests that it is not only current lead exposure that matters for a healthy pregnancy; prior exposure also matters, and it might matter more than current exposure.[42]

As can be seen in table 2.2, the mobilization of skeletal lead is correlated with the incidence of eclampsia and preeclampsia. Based on a survey of 29 preeclamptic and 101 normal pregnancies from the United States in 1997, table 2.2 reports the metal levels found in the amniotic fluid of both subsamples. At thirty-three to thirty-six weeks of gestation, preeclamptic women had lead levels that were 68 percent higher than the levels observed in the control group. At thirty-seven to forty weeks of gestation, preeclamptic women had lead levels that were 57 percent higher. The differences between the levels for preeclamptic versus normal pregnancies were much smaller for the other metals.[43]

Additional evidence indicating a connection between maternal lead levels and eclampsia can be found in a recent study of 705 pregnant women treated at three clinics in Camden, New Jersey. All of the women were eligible for Medicaid; 42 percent were African American; 19 percent were Caucasian; and 38 percent were Hispanic. Blood tests, including measurements of blood-lead levels, were conducted four times over the course of each pregnancy. Blood-lead levels increased for all women during pregnancy, but the increase was much larger in women who developed hypertension or preeclampsia. For women who developed preeclampsia the increase in blood-lead levels was twice as large as the increase in women with normal pregnancies.[44]

Table 2.2
Bone-lead mobilization and eclampsia

Metals[a]	33–36 weeks gestation					37–40 weeks gestation				
	Amniotic fluid levels					Amniotic fluid levels				
	Normal		Preeclamptic		% change	Normal		Preeclamptic		% change
	Mean	S.d.	Mean	S.d.		Mean	S.d.	Mean	S.d.	
Sodium	2.2	0.1	2.3	0.1	5	2.2	0.1	2.2	0.1	0
Potassium	217.0	24.0	227.0	24.0	5	227.0	33.0	234.0	18.0	3
Calcium	65.0	7.0	58.0	8.0	(11)[b]	66.0	7.0	62.0	8.0	(6)
Magnesium	12.6	1.5	15.6	1.0	(24)[b]	12.3	1.8	14.0	0.7	14
Iron	178.0	13.0	196.0	52.0	10	168.0	18.0	177.0	45.0	5
Copper	19.0	2.0	23.0	6.0	21	19.0	5.0	29.0	3.0	53
Zinc	2.5	1.0	1.7	0.3	(32)[b]	2.4	1.0	1.8	0.6	(25)
Selenium	10.0	1.0	7.0	0.7	(30)[b]	7.0	1.0	7.0	0.5	0
Cadmium	94.0	4.0	100.0	27.0	6	90.0	10.0	106.0	19.0	18
Lead	62.0	17.0	104.0	18.0	68[b]	58.0	26.0	91.0	28.0	57
No. of obs.	48		10			53		19		

Source: Dawson, Evans, and Nosovitch (1999).
[a] Sodium is measured in grams per liter of amniotic fluid; potassium, calcium, and magnesium are measured in milligrams per liter; iron, copper, zinc, selenium, cadmium, and lead are measured in micrograms per deciliter (μg/dl).
[b] Difference in metal levels between normal and preeclamptic pregnancies is significant at the 5 percent level or higher.

The Biochemistry of Lead Poisoning

According to one expert, over the past ten years there has been a "quantum leap" in our understanding of the "molecular mechanisms" that make lead poisonous.[45] This newfound understanding reveals that much of lead's toxicity is rooted in three biochemical and physiological bases. First, lead has the ability to mimic calcium and zinc. These two elements are essential to nearly all life forms, and are involved in the activation and mediation of several basic physiological processes. Second, a growing body of research indicates that lead alters the body's natural prooxidant/antioxidant ratio, disrupting, among other things, the human immune system.[46] Third, recent studies suggest that magnesium-dependent processes might also be a target for lead, and that lead poisoning might alter the metabolism of essential fatty acids.[47]

For adult humans, the typical body contains 1,200 grams of calcium, 99 percent of which is found in the skeleton. The remaining calcium, which is contained in the blood and soft tissue, performs four metabolic functions: It regulates cellular activity, including cell division, adhesion, and apoptosis (programmed cell death); it helps muscles contract and is critical to the transmission of nerve impulses; in the bloodstream, it neutralizes excess acidity and facilitates coagulation; and, finally, it triggers the secretion of hormones.[48]

How and why does lead compete with, and ultimately supplant, calcium in these various processes? The answer is complex, but appears to be rooted in the following observations: First, lead has the same ionic structure as calcium; both are positively charged elements with a valence of two.[49] Consequently, lead has the same chemical affinities as calcium. Second, once inside a cell, lead is absorbed by mitochondria. Mitochondrial absorption of lead disrupts calcium homeostasis and causes a buildup of calcium in the cell. This process ultimately results in premature cell death, as undue amounts of calcium in the cell trigger apoptosis. Third, because lead binds to cellular membranes at lower concentrations than calcium, it often clings to the relevant proteins and enzymes "more tightly" than calcium.[50]

Of the many calcium-related targets that lead affects, the ones that have received the greatest attention are those associated with the transmission of nerve signals. Effective neurotransmission is predicated on the cellular functions of exocytosis and endocytosis. Through exocytosis, cells release large molecules and particles, and through endocytosis other cells absorb the secreted material. Among nerve cells, lead disrupts exocytosis —the release of neurotransmitters—and it does so in an asymmetric fashion. It enhances the release of spontaneous neurotransmitters, and inhibits the release of stimulated neurotransmitters.[51] Put simply, lead exposure can cause impulse control problems, and at the same time, diminish a person's ability to respond to various social and environmental stimuli. In this regard, lead's asymmetric effects on nerve conduction might help explain the finding that lead levels are four times higher among convicted juvenile offenders than among non-delinquent high school students.[52]

Protein Kinase C (PKC) regulates a broad range of metabolic functions, including cell growth, learning, and memory. A calcium-dependent enzyme, PKC is activated by calcium. At sufficiently high concentrations, lead competes successfully with calcium for binding sites on PKC. It supplants calcium on these sites but it does not interact with the enzyme in the same way that calcium would and, as a result, it disrupts PKC's ability to trigger cell growth and regulate the processes related to learning and memory.[53] That lead is able to inhibit the release of neurotransmitters is related to PKC-associated proteins and processes. Lead interferes with the binding of the proteins synaptotagmin and syntaxin. These two proteins trigger the release of neurotransmitters in response to an inflow of calcium into nerve terminals. (Synaptotagmin is the calcium sensor that links the syntaxin-dependent fusion of neurotransmitter-laden ventricles with the nerve terminal membrane.) The connection between PKC and lead helps explain the finding that lead exposure undermines mental development and reduces IQ among children and young adults.[54]

Zinc is involved in the human growth process, fertility, and the transcription and translation of DNA. The typical adult has about 2.3 grams of zinc in his or her body, with the heaviest concentrations in the eyes, prostate, muscles, kidneys, and liver. As with calcium, one needs to ask how and why lead supplants zinc in human processes. One possibility is that lead and zinc have the same ionic structure.[55] However, there are many zinc-activated enzymes in the human body and lead appears to affect only some of them. The zinc-activated enzyme that is most sensitive to lead is δ-aminolevulinic acid dehydratase (ALAD). This enzyme has three cysteine (a non-essential amino acid) residues. One recent study suggests that lead's ability to bind to sites with three cysteines exceeds zinc's ability to bind to those same sites by a factor of 500.[56]

ALAD activates the processes that produce hemoglobin. Inside the red blood cells, hemoglobin has two related functions. It carries oxygen from the lungs to other parts of the body and, in turn, it carries carbon dioxide from the various organs and soft tissue back to the lungs. Very low levels of hemoglobin are associated with anemia, a disease characterized by weakness and a lack of energy. By supplanting zinc in the human system, lead inhibits the functioning of ALAD and concomitantly discourages the production of hemoglobin. Although full-blown anemia does not typically appear until a person's blood-lead level exceeds 40 μg/dl, there is a dose-response relationship between lead and the activity of ALAD in red blood cells. In particular, ALAD activity decreases logarithmically with increases in an individual's blood-lead level. Given this logarithmic structure, lead's incremental impact on ALAD activity is largest at low levels of exposure. For example, increasing the blood-lead level from 0 to 15 μg/dl is associated with a 50 percent reduction in ALAD activity. As ALAD activity is decreased, the concentration of its precursor molecule, aminolevulinic acid (ALA), in the bloodstream increases. Undue levels of ALA produce many of lead's symptoms, including lead colic (stomach cramps and constipation), brain swelling and pressure headaches, sleeplessness, and restlessness. Increased ALA levels might also account for some of the behavioral disorders associated with lead poisoning.[57]

Zinc plays a fundamental role in human fertility and reproduction. A crude but indicative piece of evidence in this regard is that semen contains a large amount of zinc and zinc deficiencies are associated with low sperm counts. More formally, the production and development of sperm depends on a zinc-activated protein, human protamine 2 (HP2). Displacing zinc, lead binds itself to HP2 and in the process alters the structure of the protein.[58] Zinc-related processes are also linked to the transcription and translation of human DNA. The genetic code would remain inert if not for its coupling to RNA, which carries the code, as it

is embodied in various amino acids. The code is activated through protein synthesis. Partly as a result of its affinity for zinc-dependent proteins,
lead attaches itself to RNA at certain molecular sites.[59]

That lead interacts with the transcription and translation of the genetic
code is not solely the result of its ability to mimic zinc. One recent study
suggests that lead might interfere with two early genes by activating one
form of PKC which, as explained previously, is a calcium-dependent protein.[60] Lead also affects one particular RNA molecule in a way that suggests that the molecular basis for lead toxicity might go beyond zinc- and
calcium-dependent enzymes.[61] Finally, there is evidence that lead directly
damages DNA. This evidence derives from studies showing that exposure
to lead increases urinary excretion of β-aminoisobutyric acid, a normal
degradation of thymine that is a constituent of DNA.[62]

In addition to the modes just specified, probably the most important
way in which lead affects the developing fetus is directly. Theoretically,
the placental barrier—a semipermeable membrane that separates the fetal and maternal blood streams and is composed of vascular endothelium
and other tissues—might insulate the fetus from any inorganic pathogens
carried in the maternal bloodstream. Nearly all existing studies, however, indicate that lead in the maternal bloodstream can penetrate the
placental barrier.[63] The method by which lead is transported across the
fetal barrier is not well understood. One possibility is that lead attaches
itself to certain proteins and enzymes in the maternal bloodstream that
are necessary for fetal development, and crosses the placental barrier on
the backs of such molecules. Another possibility is that the placental barrier is sufficiently permeable to allow the passage of viruses, and various
nutrients and proteins. This line of thought suggests that lead crosses the
placental barrier through a simple diffusion process.[64]

The modern-day emphasis on lead's neurotoxicity, particularly its
effects on the central nervous system of developing children, inevitably
prompts questions about lead's ability to penetrate the blood-brain barrier (BBB). The BBB is a construct that refers to the resistance most molecules confront when they spread out over the brain. Two factors inhibit
molecular diffusion in the brain. First, the capillaries of the brain have
tight cellular junctions. Second, the brain's capillaries are surrounded by
a fatty sheath composed of astrocyte cells. At sufficiently high blood-lead

levels (>80 µg/dl), lead directly undermines the integrity of the BBB and alters vascular permeability in the brain. This explains historical cases, described later in the book, in which individuals consuming extraordinarily high lead levels in their tap water eventually died from hemorrhagic convulsions and brain edema. At low blood-lead levels, lead does not appear to gain entry into the brain by directly attacking the BBB. Instead, at lower levels of exposure, lead gains entry into the brain through its ability to adhere to various proteins and enzymes, which are necessary for proper functioning of the central nervous system.[65] As already explained, lead adheres to these proteins and enzymes as the direct result of its ability to mimic calcium, zinc, and perhaps magnesium.[66]

Only recently have scientists begun to identify the mechanisms through which lead might undermine immune function. Lead attacks the immune system by altering the balance among T-cell types, which play a fundamental role in helping the human body fight off disease. Specifically, lead increases T helper 2 function, while depressing T helper 1 function. This disruption, in turn, "alters the nature and range of immune responses that can be produced, thereby influencing host susceptibility to various diseases." There is evidence that lead impairs the immune systems of the young more so than the systems of mature individuals, in much the same way as it impairs the nervous systems of the young more so than the old. One study suggests that perinatal lead exposure might increase the risk for childhood asthma.[67] Finally, lead's effects on the immune system raise the possibility that it might be carcinogenic. While animal experiments support this hypothesis, studies of human populations find little evidence that lead causes cancer.[68]

It has long been believed that lead adversely affects liver and kidney function.[69] Recent research supports this contention. As for the liver, body lead burdens are inversely correlated with the activity of a particular type of hepatic cytochrome, CYP2A6. Hepatic cytochromes are part of a system of electron-transferring enzymes in the liver essential to the organ's aerobic respiration, and the metabolic conversion and inactivation of poisons such as coumarin and nicotine. Inhibiting hepatic cytochrome activity undermines the liver's ability to take in oxygen, and in turn, create energy.[70] Lead exposure is also associated with changes in a wide variety of hepatic phosphatases, enzymes which split phosphates from organic compounds.[71] Finally, lead exposure might indirectly impair liver function by stimulating alcohol consumption. In animal experiments, rats who were not exposed to lead were averse to alcohol, while their lead-treated counterparts showed a marked increase in alcohol intake.[72]

In attacking kidney function, lead appears to work mainly through three pathways. First, lead acts directly on renal tubes. This direct effect is suggested by tubular elements that contain lead and protein.[73] Second, because lead poisoning is known to induce hypertension and hypertension is associated with renal damage, lead might indirectly contribute to kidney disease through its effects on the cardiovascular system. Third, chronic lead exposure is correlated with high levels of uric acid, a condition known as hyperuricemia. In turn, hyperuricemia causes a proliferation of smooth muscle growth in the kidney, impairing blood flow and the glomerular filtration rate. Glomerular filtration refers to the process whereby liquid and macro molecules pass through tiny semipermeable knots of renal capillaries and are eventually transformed into urine. This filtering process is based on adequate blood flow because it is pressure from the blood that forces the molecules through the complex of tiny capillaries.[74]

Many of the effects lead has on the brain, immune system, heart, kidneys, and liver are mediated through its ability to induce oxidative stress. Oxidative stress refers to the accumulation of free radicals, molecules which damage cells, proteins, and genetic material through the same chemical process that causes iron to rust.[75] A recent study of mechanics in Turkey finds that low-level lead exposure causes oxidative stress in red blood cells, and this stress likely has subclinical renal effects, including tubular damage.[76] Animal experiments reveal that oxidative stress not only damages the kidneys, but also impairs brain and liver function, and that the impact of oxidative stress on these organs varies with age.[77] In the brain, oxidative stress appears to be triggered by excess amounts of ALA, which, as explained previously, is the precursor molecule of ALAD and is produced as a result of lead's impact on the hematological system.[78] The oxidative damage wrought by lead is correlated with other clinical markers of lead poisoning, and suggests that the biomarkers of oxidative stress might also serve to identify cases of lead poisoning.[79] It is noteworthy that most researchers now believe that oxidative stress also plays a fundamental role in the onset of preeclampsia and eclampsia.[80]

Blood, Lead, and Water

The chapters that follow present evidence that water-lead levels during the nineteenth and early twentieth centuries in various parts of the world contained large amounts of lead by both historical and modern standards. This evidence is not useful unless we know the extent to which water lead was absorbed by the human system. It is certainly possible that other environmental exposures, such as industrial emissions and paint chips, were the main source of lead exposure. In light of this, a key question becomes: Is there any modern, scientific evidence showing a correlation between water-lead concentrations and blood-lead levels? If so, can one use this research to draw inferences about the contribution of water lead to body-lead burdens historically?

During the 1970s Michael R. Moore and his associates uncovered evidence that high lead levels in Glasgow's (and Scotland's in general) water supply were correlated with increased blood-lead burdens among women. At an early stage in publishing his findings, Moore's research was critiqued in two short communications to the *Lancet*. In both communications, researchers reported that when they fit linear regression models to the relationship between blood-lead levels and water-lead levels, they found that variation in water lead explained only a small portion of the variation in blood lead, and that the estimated relationship was not particularly steep.[81] This point is important as it suggests that large increases in water lead had only a small effect on blood-lead levels. Moore's subsequent research, however, revealed that the reason other scientists were finding weak relationships between water lead and blood lead was that they were imposing a linear model on a non-linear relationship. Once the underlying statistical model was respecified, a stronger and steeper connection was found.[82]

Figure 2.2 illustrates the debate between Moore and his critics. A scatter plot of the underlying raw data revealed a steep relationship between water lead and blood lead at low lead levels, but a much flatter

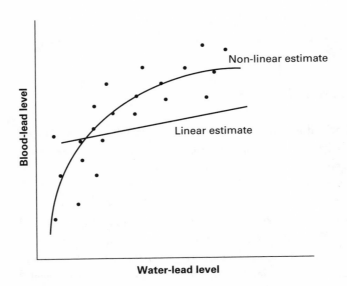

Figure 2.2
Estimating the relationship between water lead and blood lead. *Sources:* Moore et al. (1979); Moore et al. (1982).

relationship at higher levels. While figure 2.2 serves only as a convenient illustrative device, the same basic relationship is visible when one looks at the graphs originally published by Moore and his critics.[83] A non-linear model—depicted by the line labeled "non-linear estimate"—fits the raw data much better than does a linear model—depicted by the line labeled "linear estimate." Notice that the linear model understates the steepness of the relationship between water lead and blood lead at low levels of exposure, but overstates the relationship at higher levels of exposure. Moore's critics estimated the linear model when the underlying data called for a non-linear estimation.

Beyond concerns about the functional specification, there are other factors that make it difficult to correlate blood-lead levels with water-lead levels. Studies that measure the role of competing sources of lead exposure (e.g., airborne lead particles versus the amount of lead in tap water) sometimes find that non-waterborne exposure vectors explain the variation in blood-lead levels better than water-lead levels do.[84] It is easy to infer too much from such findings because exposure to water lead is subject to an error-in-variables problem. This problem introduces a downward bias into any statistical technique linking water-lead levels to blood-lead levels. There will be wide variation from house to house in the amount of tap water consumed by families, and by individuals within families. Even if a particular home has tap water with high lead levels, that will not translate into a high level of water-related exposure if people in the home drink or otherwise consume little tap water. In contrast, in a home with large amounts of lead-contaminated dust, one can be more confident that the dust will be inhaled given the universal practice of breathing. Further complicating efforts to measure the importance of competing exposure vectors is the fact that how an individual is exposed to lead will vary with age.[85]

Although the content is much lower than in the past, even today water can account for a significant portion of lead exposure. One recent study of children in Rochester, New York, found that children living in homes with water-lead concentrations greater than 0.05 parts per million had blood-lead levels that were 20 percent greater than those of children living in homes with water-lead levels below this threshold. In parts of Scotland with lead-solvent water supplies, drinking water accounts for roughly 60 percent of the population's body-lead burden. A study published in the *American Journal of Public Health* in 1989 estimated that for adults consuming water with a lead concentration of 10 micrograms per liter (just below the current EPA threshold) lead-contaminated water would account for about 7 percent of their blood lead; for children, the estimate was 14 percent. When water contains lead levels 35 times greater than the current EPA standard, it can account for about 90 percent of an adult's blood lead. Chapters 3 and 5 show that during the nineteenth century, water supplies often had lead levels 100–1,000 times greater than the current EPA standard.[86]

Summary

The key points of this chapter are that the effects of lead are multisystemic and subtle at low levels of exposure. Lead's impact on reproductive health is particularly important for later chapters. Even at low levels of maternal exposure, lead increases the risk of spontaneous abortion, stillbirths, and infant death. Although more research is needed, there is evidence that prior lead exposure is a predisposing factor for eclampsia. Lead's toxicity stems largely from its ability to mimic calcium and zinc, two elements essential to nerve conduction, reproductive health, the production of hemoglobin, and other critical physiological processes. In addition, the available evidence indicates that water-lead levels influence blood-lead levels, and that water lead can account for significant portions of a person's blood lead in some environments.

One implication from the foregoing discussion is that historical observers would have had a difficult time appreciating the dangers of lead poisoning, particularly water-related lead poisoning, because they were unable to perceive the effects of lead at a molecular level. Had they possessed such understanding and perception, perhaps municipalities would not have been so quick to embrace lead water pipes. However, the failure to recognize lead's potential as a toxin was not solely the result of an inadequate understanding of biochemistry. Some of the failure also stemmed from poor data and misspecified statistical models. Mistakes on these margins have often lead researchers to dismiss the effects of lead in particular environmental contexts, even though a closer analysis of the relevant data would have revealed a strong association between lead exposure and the associated pathology. The relationship between lead and human health is governed by complex and non-linear processes that appear random and nonexistent when researchers do not carefully control for confounding factors, allow for non-linear relationships in their statistical estimation, and design their experiments appropriately.[87]

3

Fixing Alice

In 1890, Alice worked in Leicester, England, as a machinist. She was thirty-three, married, and had four children. She was pregnant with yet another child. Perhaps her motivation was financial—she worked in an era when married women worked not by choice but out of necessity—or perhaps she just did not want another child. Whatever the case, Alice wanted an abortion. She purchased diachylon (lead plaster), rolled it into pills, and ate the pills. An abortion was induced a short time later. But Alice's homemade pills had some undesired side effects. She developed severe pains in her abdomen and extremities, jaundice, constipation, vomiting, and tremors in her hands. She entered the Leicester Infirmary on September 10, 1890, and by September 13, she had become comatose. She died soon thereafter. Alice's pills had not only terminated her pregnancy, they had also delivered her a fatal dose of lead.[1]

In 1892, Anne was twenty-two years old and worked as a hosiery hand in Leicester. She had given birth to two children during her short three-year marriage, but both had died in infancy. She had also had several miscarriages. In May 1892, Anne experienced heavy menstrual bleeding and had no return of menstruation in the following months. She had some constipation and weakness in her extremities, but no other symptoms. Then on the evening of August 8, Anne had an epileptiform convulsion. On August 9, "she lost all power in her arms and legs, and faeces were passed involuntarily." On August 10, things got even worse. There were more violent convulsions, and Anne was "continually crying out, and rolling her head from side to side." On August 11, her paralysis imploded, moving inward from her extremities to her diaphragm. She died at 8:45 a.m. that morning.[2]

Anne's doctors were baffled by her illness. Her initial symptoms were suggestive of syphilis, but her case history was not consistent with that diagnosis. Later in the course of her illness, doctors thought she might have had tuberculous meningitis or a brain tumor. Anne's autopsy, however, revealed no evidence to support such a diagnosis. "The brain was carefully examined [and] nothing abnormal" was found. All other internal organs also appeared healthy except for the intestines, which were "contracted." There was only one abnormality: On Anne's gums there was a "well-marked blue line, [which had been] inconspicuous before death." With this, doctors identified the cause of death as lead poisoning, although they were unable to say how Anne had been exposed to so much lead.[3]

At a formal inquest, Anne's aunt testified that a few weeks prior to her death, she and Anne had walked by a chemist's shop. Anne pointed to the shop and said, "That's where I get the stuff I take." When her aunt asked, "What stuff?" Anne said, "Diachylon." Her aunt was puzzled and said, "I thought that was poison." To which Anne replied, "Well, it does not poison me. I get two pennyworth and make it into pills so I can swallow them." Anne then explained to her aunt that she used the pills to induce abortions. The coroner ruled that Anne had died by taking a drug for a "felonious purpose."[4]

Single and only twenty-two years old, Mary could not remember exactly when her child was born. It was sometime in late April or early May 1898, but other than that she could not say. The pregnancy and birth had not been easy. When the child was born it could not have weighed much more than two pounds, and when it died only a month later, it weighed only two pounds, eleven ounces. Within a few days of the birth, Mary lost her sight and sense of touch. Within a few weeks, she became very weak, developed severe abdominal pains, and began having convulsions. By the time she was admitted to the Leicester Infirmary on June 7, 1898, she was confined to her bed and could not stand without assistance. During her stay in the infirmary, Mary slowly improved and was discharged three months after her admission, on September 10. By the time she was discharged, Mary had regained most of her strength, but her sight was permanently damaged and severely limited. Mary's illness and the premature death of her infant were the result of her ingestion of diachylon pills. She had started taking these pills soon after she learned she was pregnant in October 1897, and she had continued taking them until May.[5]

Married and thirty-nine years old, Charlotte thought she did not want a third child. When she became pregnant sometime in 1898, she took a two-week course of diachylon pills and induced an abortion. Initially, Charlotte experienced abdominal distress, uncontrollable vomiting, and severe pain; the symptoms soon worsened—she began having visual hallucinations and eventually fell into a state of semi-delirium. Dizzy and disoriented, she had great difficulty balancing and walking without assistance. When her doctor examined her at Nottingham General Hospital, he observed a blue gum line and immediately diagnosed Charlotte as lead poisoned. Charlotte was put on potassium iodide and magnesium sulphate, then a common treatment for lead poisoning. Charlotte improved rapidly and within a month of admission to the hospital she was released. When her physician visited her at home afterward, he found that Charlotte was in "perfect health" and was now the "proud mother" of a three-week-old baby. Charlotte's physician, however, observed that the baby was "a puny creature" who suffered from "consumption of the bowels." In his published account of the case a short time later, the physician implied that the child's small stature and poor health were the result of Charlotte's use of diachylon.[6]

The popularity of lead abortion pills grew over the years, and by the early 1900s a few unethical chemists had started mass marketing them under the guise of controlling "female problems." One popular brand of pill was described in the medical literature only as Dr. ____'s Famous Female Pills. According to the label, these pills were "world renowned and unequalled" although the label never specified for what purpose they were renowned and superior. The instructions directed the patient to take two pills four times a day. By absolute measures, these pills contained very little lead. An article in the *British Medical Journal* indicated that each pill contained only 0.0005 grains of lead, implying that if a woman took the daily dosage of eight pills, she would have been ingesting a tiny 0.004 grains a day. This did not seem like a lot until the author of the article, Dr. Arthur Hall, compared this to the level of lead found in tap water known to have caused lead poisoning. Hall cited a case in

Table 3.1
Lead in Massachusetts tap water, 1900

Municipality	Lead content of water after[a]				(Content)/(EPA level)[b]			
	Ordinary use		Standing		Ordinary use		Standing	
	Max.	Ave.	Max.	Ave.	Max.	Ave.	Max.	Ave.
Andover	0.0171	0.0108	0.0571	0.0257	11.4	7.2	38.1	17.3
Attleborough	0.1714	0.0697	0.1371	0.0905	114.3	46.5	91.4	60.3
Beverly	0.0257	0.0087	0.0314	0.0147	17.1	5.8	20.9	9.8
Bridgewater	0.0086	0.0057	0.0171	0.0143	5.7	3.8	11.4	9.5
Brookline	0.0114	0.0074	0.0286	0.0197	7.6	4.9	19.1	13.1
Cambridge	0.0086	0.0025	0.0114	0.0064	5.7	1.7	7.6	4.3
Cohasset	0.0086	0.0048	0.0086	0.0043	5.7	3.2	5.7	2.9
Dedham	0.0100	0.0082	0.0200	0.0150	6.7	5.5	13.3	10.0
Grafton	0.0229	0.0187	0.0457	0.0329	15.3	12.5	30.5	21.9
Hyde Park (new)	0.0457	0.0172	0.4571	0.0329	30.5	11.6	304.7	21.9
Hyde Park (old)	0.0200	0.0400	0.0457	0.3029	13.3	26.7	30.5	201.9
Lawrence	0.1371	0.0543	0.1829	0.0704	91.4	36.2	121.9	46.9
Lowell-Blvd.	0.0800	0.0202	0.4000	0.0861	53.3	13.5	266.7	57.4
Lowell-Cook	0.5143	0.1608	0.4643	0.2535	342.9	107.2	309.5	169.0
Metropolitan	0.0400	0.0111	0.1371	0.0293	26.7	7.4	91.4	19.5
Middleborough	0.3429	0.1549	1.143	0.6171	228.6	103.3	761.9	411.4
Needham	0.0171	0.0091	0.0429	0.0269	11.4	6.1	28.6	17.9
Newton	0.0714	0.0432	0.1714	0.0908	47.6	28.8	114.3	60.5
N. Attleboro	0.0071	0.0049	0.0329	0.0226	4.7	3.3	21.9	15.1

Webster	0.0200	0.0100	0.0571	0.0286	13.3	6.7	38.1	19.1
Wellesley	0.0152	0.0101	0.0314	0.0219	10.1	6.7	20.9	14.6
Weymouth	0.0800	0.0314	0.2286	0.1167	53.3	20.9	152.4	77.8
Mean	0.0761	0.0320	0.1705	0.0874	50.7	21.3	113.7	58.3
Median	0.0229	0.0110	0.0571	0.0290	15.3	7.3	38.1	19.3
No. of obs.	22	22	22	22	22.0	22.0	22.0	22.0

Sources: Massachusetts State Board of Health (1900), pp. 490–493; Massachusetts State Board of Health (1899).

[a] Lead content is measured as parts of lead per 100,000 units of water. Ordinary use indicates lead content of tap water after running the water for a few minutes; standing indicates lead content after allowing the water to stand in pipes overnight. The lead levels reported here were based on repeated sampling. The column labeled "Max." indicates the maximum lead level observed after sampling; the column labeled "Ave." indicates the average lead level observed after repeated sampling.

[b] Current EPA standards allow water to contain 0.0015 parts of lead per 100,000. The columns divide the lead levels observed in Massachusetts in 1900 by this modern standard.

which regular ingestion of water containing as little as 0.0028 grains of lead per gallon had produced a serious case of lead poisoning. Assuming that the patient in this case could not have consumed more than three pints of water per day, this would imply that the individual ingested about 0.001 grains of lead per day, one quarter the amount in the leaden abortion pills.[7]

Dr. Hall's comparison of the amount of lead in abortion pills to the amount of lead in drinking water was both clever and instructive. But it was also misleading. The water Hall chose contained relatively low lead levels. As will be shown, tap water in turn-of-the-century America and England routinely contained lead levels far greater than those found in a dose of Dr. ____'s Famous Female Pills.

Tap Water as an Abortifacient

In 1900, the Massachusetts State Board of Health launched an investigation into the amount of lead contained in household tap water in twenty-two municipalities across the state. Health officials took several samples of water from household faucets in these cities after the water had passed through lead service pipes, measured the lead content of these samples, and reported their findings in the annual report of the Board of Health. Officials also reported data on the chemical composition and qualities of the local water supply, including how hard the water was, and the amount of free-CO_2 (carbonic acid) it contained.[8]

Table 3.1 reports the lead levels in Massachusetts tap water around 1900, and indicates the extent to which these levels exceed current EPA guidelines. Health officials took two sets of samples, one for water following ordinary use, and another for water that was left standing in pipes overnight. There are two cities—Hyde Park and Lowell—for which there are two separate lead readings. These are given because in both Hyde Park and Lowell two separate water sources were used, and the corrosiveness of the water varied across the sources. Note that even in Cambridge and Cohasset, which had the lowest lead levels of the towns surveyed, the average amount of lead in household tap water that stood in pipes overnight exceeded the modern EPA standard by factors of 4 and 3, respectively. In the mean city, the lead level in water that had stood in pipes overnight exceeded the modern EPA standard by a factor of 58; in the median city the factor was 19. And for three cities— Hyde Park (old well), Lowell (Cook well), and Middleborough—the average lead levels in standing water exceeded current guidelines by factors of 202, 169, and 411, respectively. The discussion thus far has focused on sample averages. If one considers the observed high in each sample, lead levels exceeded the modern EPA standard by factors as large as 750.

Table 3.2 reports the amount of lead in Massachusetts tap water in terms of an abortifacient equivalent. The abortifacient equivalent refers to the amount of water an individual needed to consume in order to have been exposed to the same amount of lead as was contained in the recommended daily dose of Dr. ____'s Famous Female Pills (0.004 grains).[9] The data indicate that, in the typical town, a Massachusetts housewife would have reached the abortifacient equivalent by drinking around 80 ounces of tap water per day, assuming she regularly flushed her pipes before consuming. If, however, that housewife regularly consumed water allowed to stand in pipes for several hours, she need have only consumed 30–40 ounces of tap water per day. Furthermore, this is a situation in which measures of central tendency can be deceiving because there was a great deal of variation. Notice that in Attleborough, Lowell, and Middleborough, housewives need only have consumed 1 to 10 ounces of water daily to have reached the abortifacient equivalent. Similarly, in Lawrence, Newton, and Weymouth, housewives need only have consumed 6–28 ounces of water daily to have reached the abortifacient equivalent.

Figure 3.1 illustrates the relationship between the abortifacient equivalent and the modern EPA standard regarding water lead. The y-axis is scaled logarithmically and the function approaches both axes asymptotically. The figure shows that when water-lead levels exceed the modern EPA standard by more than a factor of 300, consuming less than 2 ounces of water per day would allow one to reach the abortifacient equivalent. In contrast, when water-lead levels are only 2–4 times greater than the modern EPA standard, one need consume between 150 and 300 ounces of water per day to reach the abortifacient equivalent.

Tables 3.1 and 3.2 indicate that the lead levels in Massachusetts tap water varied greatly from town to town. The primary source of this

Table 3.2
The abortifacient equivalent in Massachusetts water

| Municipality | Ounces of water needed to reach abortifacient equivalent | | | |
| | After ordinary use | | Standing water | |
	Maximum	Average	Maximum	Average
Andover	51.4	81.3	15.4	34.2
Attleborough	5.1	12.6	6.4	9.7
Beverly	34.2	100.9	28.0	59.7
Bridgewater	102.1	154.1	51.4	61.4
Brookline	77.0	118.7	30.7	44.6
Cambridge	102.1	351.3	77.0	137.2
Cohasset	102.1	183.0	102.1	204.2
Dedham	87.8	107.1	43.9	58.5
Franklin	—	30.7	—	7.7
Grafton	38.4	47.0	19.2	26.7
Hyde Park-new	19.2	51.1	1.9	26.7
Hyde Park-old	43.9	22.0	19.2	2.9
Lawrence	6.4	16.2	4.8	12.5
Lowell-Blvd.	11.0	43.5	2.2	10.2
Lowell-Cook	1.7	5.5	1.9	3.5
Marblehead	—	102.1	—	61.4
Metropolitan	22.0	79.1	6.4	30.0
Middleborough	2.6	5.7	0.8	1.4
Needham	51.4	96.5	20.5	32.6
Newton	12.3	20.3	5.1	9.7
North Attleboro	123.7	179.2	26.7	38.9
Norwood	—	204.2	—	6.4
Webster	43.9	87.8	15.4	30.7
Wellesley	57.8	87.0	28.0	40.1
Weymouth	11.0	28.0	3.8	7.5
Mean	40.3	88.6	20.4	38.3
Median	41.1	81.3	17.3	30.0

Sources: Massachusetts State Board of Health (1900), pp. 490–493; Hall (1905).
Note: The abortifacient equivalent (AE) can be expressed as

$$AE = 0.878/L,$$

where L equals the water-lead concentration expressed as parts per 100,000. The derivation of this equation can be broken down into the following three steps:

Table 3.2
(continued)

1. Lead levels in Massachusetts were originally reported as parts per 100,000. These measures had to be converted to grains per (U.S.) gallon by the following formula:

$y = (0.583) \times (L)$,

where y equals the lead level expressed as grains per gallon and L is defined as above.

2. To arrive at the fraction of a gallon of water necessary to ingest the abortifacient-equivalent, the following formula was used:

$F = (0.004)/y$

where F equals the fraction of a (U.S.) gallon of water that would have contained the same amount of lead as the recommended daily dose of Dr. ____'s Famous Female Pills; and y is the amount of lead in the water measured as grains per gallon. The value of 0.004 in the numerator reflects the fact that the recommended dosage (eight pills) contained 0.004 grains of lead. For the intuition behind this equation, note the following. If $y = 0.008$, then an individual would have had to consume one-half gallon of water to get the abortifacient-equivalent; if $y = 0.004$, then one gallon; and if $y = 0.002$, then two gallons.

3. Because an ounce equals (1/128) of a gallon, the ounces of water that would have had to have been consumed to reach the abortifacient-equivalent (AE) in ounces was calculated as follows:

$AE = (F) \times (128)$.

The equation $AE = 0.878/L$ follows by substitution.

variation appears to have been water softness. Water supplies that were soft tended to be more corrosive and absorb more lead from the interior of service pipes than water supplies that were hard. Why does water softness affect the lead solvency of water supplies? Hard water contains high levels of calcium and magnesium, which help neutralize the acids that can form in water, and promote the formation of a protective coating on the interior of pipes. Soft water contains relatively low levels of calcium and magnesium. (Chapter 6 explores the factors that influence a water supply's lead solvency in greater detail.)[10]

The age of lead water lines also affected how much lead was dissolved into drinking water. Over time, a protective coating formed on the interior of lead water lines and this limited how much lead leached into the water. In short, holding the corrosiveness of local water supplies

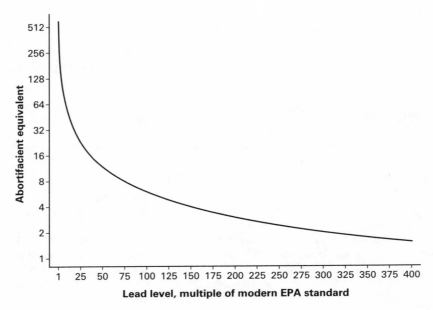

Figure 3.1
Abortifacient equivalent and modern EPA standard. *Sources:* See notes to table 3.2. for derivation of abortifacient equivalent; Hall (1905). For lead levels in Massachusetts tap water, see Massachusetts State Board of Health (1900), pp. 490–493. *Note:* The y-axis has a logarithmic scale.

constant, systems with old lead pipes exposed consumers to much less lead than systems with new lead pipes. The influence of pipe vintage can be illustrated by both experimental evidence and historical observation. For example, in 1936 two scientists in New York City compared the amount of lead that leached out of old lead pipes to that of new pipes. They found that substantially more lead was introduced into water stored in new pipes than in old pipes.[11] Studies conducted by the Massachusetts State Board of Health during the 1890s and early 1900s found that the most serious cases of lead poisoning tended to occur with newer lead pipes as opposed to old pipes.[12]

What were the effects of all this lead on fetal and neonatal development? That lead levels in water were comparable to those found in black-market abortifacients suggests the effects were severe. To provide

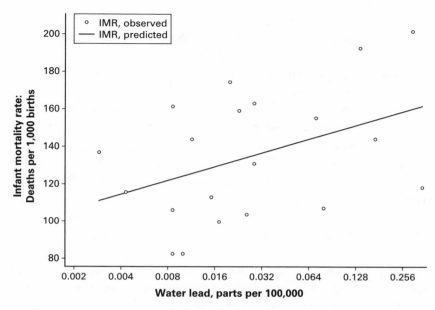

Figure 3.2
Water lead and infant mortality in Massachusetts, 1900. *Sources:* Massachusetts
State Board of Health (1900), pp. 490–493; Troesken (2006b); Massachusetts
State Board of Health (1899). See also appendix A for data on births. *Note:* The
R^2 on the trend line is 0.176, with an estimated slope of 10.618, which is signifi-
cant at the 0.033 level (one-tailed test).

a more concrete picture, figure 3.2 plots the infant mortality rate against
the observed water lead level in each of the towns reported in table 3.1.
A trend line is also plotted. Figure 3.2 suggests a positive and log-linear
relationship between water lead levels and the infant mortality rate, such
that at low lead levels (those around 0.002 parts per 100,000) the infant
mortality rate averaged around 110 infant deaths per 1,000 live births,
while at the highest lead levels (those at 0.2 and above) the infant mortality rate averaged 160. This implies that infant mortality rates in those
cities with the highest water lead levels were 45 percent higher than those
in cities with the lowest levels.

A more formal and extensive econometric analysis of the relationship
between lead water pipes and infant mortality is offered in appendix A.
This analysis includes a larger sample of cities from both Massachusetts

and England, and subjects the data to a battery of tests. The upshot of this analysis is that in cities using lead water pipes, infant mortality and stillbirth rates were between 8 and 25 percent higher than in cities using non-leaden pipes, depending on region, time period, and estimating procedure. Furthermore, most of the difference between lead and non-lead cities was driven by the experience of cities with new lead pipes or extremely soft water. In these places, infant mortality and stillbirth rates were more than 50 percent higher than in cities without lead pipes. Finally, the basic finding that lead water pipes was associated with elevated infant and fetal mortality is robust; it survives all reasonable changes in econometric specification.

How Come No One Noticed?

The statistical and qualitative evidence presented thus far prompts the question: If lead water pipes were having such severe effects on fetal and infant health outcomes, how come no one noticed? The answer has two interrelated parts. First and foremost, some people did notice. The only problem was that it took decades for those people to accumulate evidence linking lead water pipes with adverse fetal and infant health outcomes and, even then, the evidence was not definitive. Second, unaware or skeptical of the dangers lead-contaminated water might pose for the developing fetus, most scientists focused on how water lead affected the health outcomes of older children and adults. Because these groups were much less vulnerable to the effects of lead than the developing fetus, their health was a poor barometer of the effects of high water-lead levels.

Probably the first person to suggest a connection between lead water pipes and poor reproductive health outcomes was Alfred Swann. In 1889, Swann published a short article in the *British Medical Journal*, documenting the case histories of three women, Mrs. T, Mrs. C, and Mrs. F. For many years, Mrs. T had suffered from colic, rheumatism, irregular menstruation, and, periodically, severe menstrual hemorrhages. She also could never bring a child to term. In one year alone, Mrs. T had two miscarriages. Mrs. C had similar symptoms, though she had aborted only once, and was otherwise unable to conceive. Mrs. F had had two children before developing symptoms of colic, constipation, and an irregular menstrual cycle. She also lost her ability to bring a child to term.[13]

In each of these cases, Swann eventually diagnosed the women with lead poisoning and recommended that they replace their lead water pipes with pipes made of iron. Despite having husbands who were sometimes skeptical of the notion that lead water pipes could have had such serious consequences, the women consented to this request and their health rapidly improved, as did their reproductive efforts. All three women became pregnant and delivered healthy children once the lead piping was removed from their homes.[14]

Three years later, in 1892, Swann published a second article on the dangers of lead water pipes. Titled "A National Danger: Lead Poisoning from Service Pipes" and appearing in the *Lancet*, the article made the case that Britain had failed to adequately appreciate, and respond to, the dangers of lead water pipes. Many "years have now passed," Swann wrote, "since I first drew attention to the great danger of supplying water for drinking and dietetic purposes through leaden pipes." But, unfortunately, as far as Swann could gather, "the really fearful danger existing from this cause has either been ignored or only partially recognised." Policymakers and the populace were apathetic and the doctor wanted to change that: "The general opinion seems to have been that the results of drinking water very slightly contaminated by lead were inconvenient rather than dangerous, and my object in writing the present paper is to show that a serious national danger exists."[15]

Swann found it ironic that his contemporaries were oblivious to the dangers of lead water pipes, while professionals in the ancient world condemned or banned the use of lead pipes. "When we realize that half a century before the Christian era Vitruvius condemned lead service pipes and 130 years after that time Galen did the same thing," Swann wrote, "it is a puzzle to my mind to know how it is that the use of such a poisonous means of conveying water should ever have become general." The suffering caused by lead-contaminated water left Swann nostalgic for the ancient world. He was particularly concerned about the "children begotten by [lead-]poisoned parents"—children who, by all accounts, were "puny, rickety, and ill developed, both mentally and physically."[16]

Why did Swann fail to convince the medical community, as well as the broader society, that lead pipes posed a serious health risk? As explained in chapter 2, it has been only in the last ten years that medical researchers have uncovered systematic evidence that continual low-grade lead exposure (as occurs with water) increases the risk of spontaneous abortion, stillbirth, and infant death. However, at the time Swann wrote his articles (1889 and 1892), almost all of the evidence linking maternal lead exposure to fetal and infant death focused on occupationally exposed women.[17] Because occupational exposure typically involved much higher lead levels than water-related exposure, Swann had only a scant scientific literature on which to base his claim. The fact that high levels of lead exposure induced abortions and premature births did not mean low levels of lead had the same effect.

Evidence in favor of Swann's position developed slowly over the course of the early twentieth century. It was more than twenty years later that the *British Medical Journal* published another article exploring the links between lead water pipes and reproductive health outcomes. In this article, W. W. Stainthorpe explored an epidemic of lead poisoning in Guisborough, England, which was traced back to the city's water supply. Stainthorpe's article was "confined to 120 serious and typical cases" from the epidemic about which he "had made careful notes." All of the fifty-five women in Stainthorpe's sample who were of childbearing age experienced menstrual disorders. "The majority of female patients suffered from profuse menorrhagia [excessive menstrual flow] lasting in some instances for a period of three weeks." Five women reported spontaneous abortions; one of these had aborted three times over a five-year period; another aborted twice in three years.[18]

Seventeen years later, in 1931, Ernest Milligan published a short communication in the *British Medical Journal* revealing further evidence. Milligan observed that "during the last year the infant mortality rate in an area [supplied] with a very lead-soluble water was 134, while that of an area in the same district in which the water had been treated to prevent lead solubility was 56." Milligan noted that in the area with lead-soluble water, infantile convulsions and prematurity were prominent causes of death. In concluding his article, Milligan tried to draw a general lesson from this simple comparative exercise, writing, "In those districts with a high [infant mortality, the water] supply should be examined for the presence of lead."[19]

For skeptical observers, however, the evidence accumulated by Swann, Stainthorpe, and Milligan was still insufficient. There were three areas of concern. First, all three authors presented their evidence in the form of uncontrolled case studies. As a result, it was difficult to know whether it was water lead, or some other confounding variable causing the adverse health outcome.[20] Second, many observers were skeptical of the proposition that lead could travel from the mother's body to the developing fetus. It was possible to address this concern by studying lead levels in the placenta and through the autopsies of stillborn babies and those who died shortly after birth. Unfortunately, this sort of evidence was sparse and often dismissed as inconclusive.[21] Third, neither Swann nor those who followed him could explain the mechanisms through which lead harmed the fetus. Without a clear understanding of this process, there would always be doubt surrounding the proposition that lead in general, and water lead in particular, adversely affected fetal development.[22]

It was only with a series of animal experiments conducted during the 1920s and 1930s that researchers began to appreciate lead's ability to travel from mother to unborn child. In 1925, a group of British researchers published an intriguing paper exploring the effects of lead, copper, thallium, and thorium on the development of the rabbit fetus. They showed that within a particular window of lead exposure, they could induce abortion of the fetus without causing any obvious effects on the mother. Anticipating late-twentieth-century studies, these researchers also showed that lead could penetrate, and affect, the placental barrier. The other metals studied by these researchers had no systematic effect on the risk of abortion. In another study using rabbits, a Japanese researcher demonstrated that lead could travel from the mother to the fetus, and that the effects on the fetus were pervasive.[23]

There was also evidence from the natural world that seemed to support the supposition that continually ingesting small amounts of lead could adversely affect reproductive health. An early-twentieth-century study by E. Morgan examined the pregnancy outcomes of sheep from a Welsh area dominated by lead mining. The grass on which the sheep grazed was contaminated by small amounts of lead. Morgan found that the sheep aborted often, and gave birth to offspring that were "under-sized, delicate, and rarely survived" more than a few weeks. However, once the sheep were moved and began consuming grass that was not polluted with lead, the abortions stopped and the animals had healthy offspring. Along similar lines, K. Carpenter found that dogs and cats in lead-polluted areas failed to breed, and that younger animals died prematurely. She also documented the case of an apparently healthy mare in a lead-polluted area that gave birth to five foals over a twelve-year period. Each of the foals died from plumbism within five years of birth.[24]

Eugenics and Lead Poisoning

Without a well-developed body of scientific evidence on the effects of lead on fetal development, reform-minded physicians sometimes turned to unscientific or pseudo-scientific theories to bolster their claims. In the years between 1880 and 1910, Sir Thomas Oliver was arguably the world's foremost authority on lead poisoning. Beyond his scientific work, Oliver was also a forceful advocate of industrial reform and various public health initiatives, including the passage of legislation to protect workers from undue lead exposure, the regulation of lead-based paints, and the elimination of lead water pipes. According to Oliver, all of society had a stake in preventing lead poisoning, particularly among children and fecund women. Lead, he claimed, brought about the "physical degeneracy" of future generations and prevented women from rising "to the dignity of the completed act of motherhood." This, in turn, would undermine the long-term viability of the human race, not just in England but worldwide. Simply put, Oliver believed lead was a race poison.[25]

On May 4, 1911, Oliver presented a paper at the Eugenics Education Society in London, England. The paper was titled, "Lead Poisoning and the Race," and versions of it were later published in both the *British Medical Journal* and the *Eugenics Review*. Oliver framed his paper using the language and pseudo-scientific constructs of the eugenics movement. "From a physical point of view," he wrote, "a nation's strength is measured by its reproductive power and the high percentage of the fitness of its children." "Good physique," adequate "numbers," and "brain power" were essential for British victory in "future wars." In light of this, it was "not before time" to ask "for some explanation of" England's "declining birth-rate" and the high rate of abortion in the country: According to Oliver "one of every six pregnancies" ended "in abortion." Although Oliver did not state it explicitly, he implied that a decline in the quality of newborn children went hand-in-hand with these trends.[26]

What were the forces behind the declining birth rate and the excessive rate of spontaneous abortion? According to Oliver, one force was the "modern craze for pleasure" that was "undermining the best instincts of motherhood [among all] classes of society." Another force was the general practice of employing women in manufacturing. Too many pregnant and nursing women worked when they should have spared themselves, Oliver argued. "Hard and exhausting work followed till towards the end of pregnancy" not only caused "infantile immaturity," but also reduced the resistance of the expectant mother to the fatigues incidental to the stages of childbearing, and produced "an ill-nourished condition of the infant when born."[27]

For Oliver, though, the most important factor behind the decline of Britain's human stock was lead. According to Oliver, lead hit "hard the reproductive powers of man and woman, but especially of woman." Lead destroyed "the developing life by directly poisoning it, or it check[ed] the growth of the foetus in the womb by cutting off its channel of nutrition." To show that lead harmed the developing fetus, Oliver presented statistics from industrial employment which showed that women working in lead-related industries had much higher rates of stillbirths and miscarriages than did women employed in non-lead industries; he discussed the results of animal experiments where maternal lead exposure induced abortion; and he offered a brief description of an epidemic of stillbirths in one Yorkshire town. Upon investigation, medical officials traced the epidemic to the town's use of lead water pipes, and once this cause was removed, the epidemic subsided and the rate of stillbirths declined.[28]

In addition to inducing stillbirth and abortion, Oliver speculated that lead might also have had an effect on the children who survived lead exposure in utero. This channel was perhaps the most important for those who viewed the world through the lens of eugenics: "One question which the Eugenics Education Society cannot but feel interested in, as regards to lead-poisoned parents, is, what are the possible effects of lead upon the children who survive?" Oliver acknowledged that the existing evidence indicated that there were no long-term effects of in utero lead exposure, but he maintained that this was "contrary to expectation." He then drew parallels to the children born to syphilitics and alcoholics. While children born to syphilitic mothers sometimes survived and grew up to "appear quite healthy," Oliver pointed out that "many asylum physicians" believed that "general paralysis in middle-aged men [was] in many instances the result of congenital syphilis." Similar was the case of infants born to alcoholic mothers who "frequently [died] of malnutrition weeks after birth"; those that survived then exhibited "in later years an instability of the nervous system" and a greater susceptibility to alcohol intoxication. The same sort of patterns, Oliver hypothesized, might occur with lead as they did with syphilis and alcohol.[29]

Eugenicists argued that immigration and intermarriage allowed undesirable groups to infiltrate society and disrupt the transmission of "positive" human characteristics. According to Oliver, lead worked similarly, disrupting natural and healthy modes of human reproduction. "It is difficult to say," Oliver wrote, "to what extent plumbism is affecting the future of the [human] race, but already in Hungary, and less in Staffordshire, there are signs that the development of child-life is to some extent being interfered with." Given the toxicity of lead and its powerful effect on human reproduction, Oliver doubted that "any life brought into existence" under its influence could have withstood the pernicious effects of a metal "so stereotyped into its tissues."[30] In the absence of state intervention and regulation of exposure at work and at home, society could only watch as the effects of lead were passed from one generation to another.

A long-run perspective suggests that Oliver did himself and his cause a disservice by using the model of eugenics to dramatize the intergenerational effects of lead poisoning. The scientific evidence that lead might interfere with reproduction and have a persistent effect across generations is quite strong; the same cannot be said of eugenics. Oliver's argument that widespread lead exposure would undermine future generations was correct; the eugenicist's argument that bad genes, allowed to multiply unchecked, would doom future generations was not. Furthermore, a growing body of research suggests that Oliver's characterization of lead as a race poison was, if anything, overly narrow. There is much evidence to suggest that a broad range of environmental contaminants might currently be poisoning future generations.[31] A short-run perspective, however, suggests Oliver's foray into eugenics might have served a positive end. By employing eugenics concepts, Oliver was able to put his arguments into a construct that gave meaning and coherence to the myriad of facts and statistics that surrounded lead poisoning.

Although it is impossible to know for certain why Oliver chose to use the rhetoric of eugenics, one distinct possibility is that it was an act of intellectual desperation. In an age when the effects of lead were so poorly understood, plumbism appeared a pedestrian, and relatively minor, illness. As explained in later chapters, most people in the nineteenth century equated water plumbism with rheumatism and colic, and these pathologies appeared minor when compared to the ravages of cholera, typhoid fever, and smallpox. There were, it appeared, other more pressing health issues to address. Having studied lead his entire career, Oliver not only sensed, but knew from years of first-hand observation, that lead's effects on the human system, while often subtle, were no less significant than were the effects of more overtly terrifying diseases.

But whatever Oliver's motivations, one thing is now clear. There is no longer any need to appeal to the principles of a discredited pseudoscience. The last twenty years of scientific research have identified the many ways lead poisons people, both upon exposure and much later. Lead was, and continues to be, a race poison—if by race, one means the broad swath of humanity and not a few privileged groups. Lead exposure has intergenerational effects. As discussed in chapter 2, women and men who were lead poisoned as children but who were not exposed as adults have increased rates of miscarriages and abortions and have children who are more likely to exhibit developmental delays. Similarly, women who were exposed to undue lead levels in their drinking water before and during pregnancy had higher rates of spontaneous abortion and infertility. As shown in appendix A, the use of lead water pipes around 1900 increased stillbirth and infant mortality rates in England and Massachusetts by between 8 and 25 percent.

A Skeptical Calculus

The tendency for nineteenth-century scientists to focus on adult health in predicting the safety of lead water pipes is well illustrated by Professor Eben Horsford. The City of Boston consulted Horsford during the 1840s when it was first considering the use of lead in its water distribution system. In an ingenious series of calculations, Horsford estimated how much Boston tap water an adult would have had to consume in order to die of lead poisoning. Horsford began by citing medical evidence indicating that an individual would have to accumulate 3.38 grams of lead in his bodily tissues before succumbing to lead poisoning. He also assumed that two-thirds of all lead consumed through drinking water would be retained in the body, a generous assumption that favored his opponents.

With these data in hand, Horsford estimated that an individual would have to drink 28,194 gallons of Boston tap water before succumbing to lead poisoning. Making the extreme assumption that an individual drank a gallon of water a day and never flushed his pipes to clear out the lead, it would have taken seventy-seven years for a person to consume that much water. Under more realistic assumptions about consumption, Horsford suggested that an adult could consume Boston tap water for 3,465 years before developing fatal plumbism.[32]

With the benefit of some hundred years of hindsight, it would be easy to quibble with the specifics of these calculations. Consider again Horsford's estimated time-'til-death numbers. His assumption that 5 grams of lead was the minimum dose it would normally take to induce fatal lead poisoning was probably off by a factor of 10; ingesting one-half a gram of lead is normally fatal. Horsford also inadvertently used incorrect information in calculating the amount of lead in Boston tap water. The professor assumed that Boston tap water contained between 0.016 and 0.048 ppm. This was far too low; estimates reported by the Massachusetts State Board of Health in 1899 indicated that lead levels in Boston tap water averaged 0.29 ppm, and in some cases were as high as 1.37 ppm.[33] Adjusting Horsford's calculations accordingly would suggest that an individual could have become lead poisoned in a much shorter time span, perhaps in only a few years' time.

Probably the strongest evidence that Horsford's calculus was wrong was Boston's subsequent experience with lead water pipes. In an article published in the *Boston Medical and Surgical Journal* in 1889, Dr. E. M. Greene documented two serious cases of lead poisoning caused by Boston's public water supply. The first case involved a family living on Bennett Street. The mother "enjoyed fair health," while the father and son "suffered from general debility and marked anaemia." The son, who appears to have been in his late teens or early twenties, "had been unable to work for several months, and suffered from severe, frequent, abdominal colic, gastro-intestinal disturbances and general muscular weakness."[34]

Dr. Greene's second case involved a middle-aged mother and two adult daughters residing on Hancock Street. "The mother and younger daughter were very anaemic and easily fatigued." The mother suffered from "numbness," and at times, "a complete loss of sensation" in her arms. The younger daughter had "for a long time [been] subject to severe muscular cramps of short duration and sudden onset, affecting mostly the calves, but often, also, the muscles of the trunk." The elder daughter had "for several years . . . suffered from severe attacks of abdominal colic, attended often with vomiting and diarrhea, and followed by constipation." Over the years, these attacks grew increasingly "frequent and severe in spite of medicine to regulate the bowels and the most careful attention to diet."

Only when they discontinued the consumption of Boston tap water did Greene's patients show improvement. Moreover, the tap water from the patients' homes, as well as homes chosen randomly from around the city, showed unusually high lead levels even by the standards of the nineteenth century. One study found that Boston tap water contained lead levels as high as 0.1–0.2 parts per 100,000, or 67–133 times the current EPA standard.[35] There is also evidence that, at this time, up to 40 percent of the Boston population had elevated levels of lead in their urine.[36]

But whether or not one accepts Horsford's calculations as plausible, there are two larger issues here. First, lead levels need not have reached the point where they were actually killing people to have been unsafe. As the foregoing cases suggest, even doses of lead that were not fatal were capable of causing paralysis, colic, sleeplessness, fatigue, headaches, and so on. Second, Horsford studied the effects of lead on the wrong population, as did most other officials during this period. Rather than focusing on adult health outcomes, he and others should have been looking at the impact of lead pipes on fetal and infant development. It was on this group that lead water pipes caused the greatest damage. If lead levels reached the point that they were causing paralysis among adults, it is not difficult to imagine what they were doing to the unborn and the very young. Even fairly minor adult symptoms, such as constipation and lethargy, likely suggested more serious fetal and childhood injury.

Although Horsford's calculations appear to have been incorrect, the logic behind them was compelling and would be used repeatedly in the years that followed. For example, in their textbook published in 1925, John C. Thresh and John F. Beale used a calculus very similar to that employed by Horsford more than a half-century earlier to establish a safety threshold for water lead. Thresh and Beale assumed that the typical person consumed three pints of water per day, or 137 gallons of water per year. Given this assumption, "to ingest 1 grain of lead from water containing 1/100 grain per gallon, the consumer would have to drink at this rate for nine months." Like Professor Horsford, Thresh and Beale believed that a healthy person would have easily eliminated such small amounts of lead from his or her system.[37]

Building on this calculation, Thresh and Beale went on to speculate that "a water containing" as much as "1 part of lead in one million" would have been "quite safe." They conceded that there was "no evidence" that such was the case "at the time", but that it was probable. To put this in context, the threshold lead-level recommended by Thresh and Beale would have exceeded the modern EPA standard by a factor of 66.7. One need only have consumed ten ounces of such water every day to have been ingesting the same amount of lead as was contained in the daily dose of Dr. ____'s Famous Female Pills. Furthermore, Thresh and Beale were not the sort of people who dismissed the possibility of water-related lead poisoning out of hand. On the contrary, both writers had previously published articles in the *British Medical Journal* documenting cases of water-related lead poisoning.[38] That two scientists so attuned to water lead recommended a standard as high as 1 ppm illustrates the propensity for, and significance of, using adult health to benchmark safety thresholds.

A Best-Case Scenario

What were the practical consequences of using adult health outcomes to gauge lead levels in local water supplies? Misdirected by this approach, how long did it take local officials to discover they had a water lead problem? Were officials reluctant to respond if there were only a handful of adults and older children made ill? More importantly, what did this approach imply for the health outcomes of pregnant women and their unborn children? To answer these questions, the following section describes how the water company in Milford, Massachusetts responded to discoveries of excess lead in its water supply. What follows is a best-case scenario; the water company in question responded constructively as soon as it became aware of lead-related illnesses among adults and older children. (Chapter 8 presents a worst-case scenario.)

The Milford Water Company was created in 1881. It provided water to two neighboring towns on the Charles River, Milford and Hopedale. Initially, the company drew its water from three wells along the river, just north of Milford. Although this water was lead solvent, and the company used lead service pipes to connect residences to street mains, the company was not made aware of any cases of water-related lead poisoning until an investigation conducted by the Massachusetts State Board of Health in 1897. After consulting nine physicians in the Milford-Hopedale area, the Board of Health found that there were sixteen adults suffering from mild to moderate cases of lead poisoning, apparently of unknown origin. The Board of Health then analyzed the tap water in the homes of these individuals. All told, twelve homes were tested for water-lead levels, and the findings left little doubt that lead-contaminated water was the source of the lead poisoning. In one home, the water-lead level was 14.5 ppm, 969 times the modern EPA standard. In another home, the water-lead level was 11.6 ppm, 775 times the modern EPA standard.[39]

In the five years after this discovery, the Milford Water Company took several steps to reduce the lead solvency of its water supply. The company began to remove the lead service pipes and replace them with pipes made of safer materials. This, however, was a slow process because replacement was expensive. The company also abandoned the wells it had been using as the city's water source, and began drawing water from the

Charles River. Water from the river was more polluted than that from the wells, and therefore had to be filtered. The company installed a slow-sand filtration system, which not only eliminated bacterial contaminants, it also reduced the lead solvency of the river water.[40] The introduction of these measures in 1902 and 1903 reduced the average water-lead levels in area homes from 1.39 ppm (627 times the modern EPA standard) to 0.27 ppm (175 times the modern EPA standard). Although 0.27 ppm is a high lead level by modern standards, it was well below the 0.5-ppm threshold then considered safe by the Massachusetts Board of Health.[41]

After introducing a new water supply and adopting slow-sand filtration, the Milford Water Company "believed that it had done everything necessary to protect its consumers" from lead poisoning. But in 1914, a decade after these precautions had been adopted, "a suspicious case of illness occurred" in the home of an individual who employed an unusually long lead service pipe. Although the water company believed that this individual had an "unusually sensitive" constitution, it conducted tests on the individual's tap water and found high lead levels. The company began searching for an appropriate chemical treatment that would reduce water-lead levels even further. Engineers for the company found that while the addition of chalk or magnesia significantly reduced the lead solvency of the water, the most effective agent was quick lime. Within a few months of discovering this one mild case of lead poisoning, the Milford Water Company had constructed, and put into operation, a neutralizing plant.[42]

The introduction of lime dosing in 1914 had an immediate and beneficial effect on water-lead levels in the area. In particular, the average water-lead level in Milford and Hopedale fell from 0.27 to 0.1 ppm. Although the 0.1 level exceeds the modern EPA standard by a factor between 6 and 7, it was well below the threshold considered safe around 1910 and 1920.[43]

There are two notable features of Milford's experience. First, by the standards of the day, Milford acted quickly. As soon as officials discovered the lead in their water was making people sick, they adopted measures that reduced the lead solvency of the water. As will be made clear in later chapters, Milford accomplished in ten to twenty years what it took Glasgow and New York City more than a century to accomplish.[44] Second, from 1881 through 1897, consumers in Milford were consuming water with lead levels 600 times greater than the modern EPA standard, yet no one in the city was aware there was a problem until state health officials investigated. With lead levels this high, consumers need only have consumed 1 ounce of tap water per day to have reached the abortifacient equivalent. This situation persisted for sixteen years. After 1903, consumers in Milford consumed water with lead levels 175 times greater than the modern EPA standard, yet no one knew there was a problem until a person with an "unusually sensitive" constitution became ill. With lead levels this high, consumers need only have consumed 4 ounces of tap water per day to reach the abortifacient equivalent. This situation persisted for ten years.

Summary

Water-lead levels in turn-of-the-century Massachusetts were high by modern standards, often exceeding the current EPA threshold by 100 times or more. Lead levels in some towns were so high that drinking 10–20 ounces of tap water per day would have been equivalent to consuming the recommended daily dosage of black-market abortion pills. Statistical evidence suggests that the use of lead water pipes in Massachusetts and the north of England increased stillbirth and infant mortality rates by 8–25 percent. Effects were even larger in places with new pipes and/or with highly corrosive water supplies. Historical observers typically could not detect these effects because they focused on adult health outcomes in their efforts to assess the dangers of lead water pipes. Because adult health was such a poor barometer of water lead levels, consumers often had to tolerate the problem for two to three decades in even the best-case scenarios. Although a few scientists like Alfred Swann recognized the dangers of lead pipes, scientific evidence on this issue accumulated slowly over the course of the late nineteenth and early twentieth centuries.

4

The Latent History of Eclampsia

Only nineteen years old and eight-and-a-half-months pregnant, Eileen was diagnosed with eclampsia on November 15, 1893. Her symptoms appeared consistent with that diagnosis. Her temperature was slightly elevated; her pulse was rapid and suggested hypertension; her urine contained a "copious" amount of albumen, an indicator of kidney trouble. Eileen was also very drowsy and had experienced a violent seizure. If swift steps were not taken, there was the risk she might develop nephritis and her unborn child might succumb to the toxins her body was producing. Eileen was quickly admitted into the Sheffield Union Infirmary where the attending physicians launched an aggressive intervention. They gave her potassium bromide every four hours in an effort to mitigate any kidney damage, laxatives, and diaphoretics to make her perspire. Her symptoms, however, worsened and she soon slipped into a state of semi-consciousness.[1]

On November 17, at 2:30 p.m., the convulsions returned. Her "eyelids blinked violently," her "limbs became rigid," and her face turned blue as she struggled to breathe. All the while, her jaws "were firmly clenched." At 2:50, while being transferred to the maternity ward, Eileen had a second fit. Doctors administered chloroform to stop the convulsions, and glycerine to induce labor. The third seizure began at 4 p.m., a half-hour after the chloroform was stopped. The fourth fit was at 4:30, the fifth at 5:00, the sixth at 5:20. The seventh and "most severe" episode began at 6:20 p.m. Chloroform was again administered and she was forcibly dilated. Eileen's child was removed by forceps over the course of the next twenty minutes. Soon after birth, the child stopped breathing but quickly recovered. Three days later, on November 20, Eileen regained consciousness. Although the preceding days were "a complete blank to her," on November 24, she was "very lively" and said "she felt quite well."[2]

When Eileen was discharged from the hospital on December 13, 1893, her urine was devoid of albumen and her kidneys were apparently free of disease.[3] In some sense, Eileen was lucky: Between 17 and 30 percent of all expectant mothers stricken with eclampsia perished as a result. Yet Eileen's long-term prospects were not promising. According to one study, 40 percent of all eclampsia survivors in England and Wales in the year 1930 suffered from an eclampsia-related disability such as "chronic invalidism," heart disease, hypertension, or severe anemia. Another study indicated that 33–40 percent of all survivors died from renal failure within a few years of recovery. In short, even when a woman survived eclampsia in the short term, her long-term health prospects were compromised.[4]

There is no record of what ultimately became of Eileen's child. It is possible that the child survived the birth process without any permanent damage. It is also possible that the toxin that had attacked the mother also attacked the child. The medical literature of the late nineteenth century contains several references to children born to eclamptic mothers who eventually developed convulsions or nephritis and perished shortly after birth. Because mother and child usually presented very similar or identical symptoms (e.g., convulsions and kidney failure), many physicians believed that they were killed or injured by the same toxins. The odds that Eileen's child escaped the birth process alive and in good health are not heartening. Today, 12 percent of all children born to eclamptic mothers die within the first few months of life. At the turn of the twentieth century, as many as 50 percent of all children born to eclamptic mothers died soon after birth.[5]

Theories about Eclampsia

Doctors in 1893 had many theories about what caused women like Eileen to develop eclampsia. Some physicians theorized that eclampsia might be related to prior exposure to infectious diseases. Others believed that a particular viral or bacterial pathogen caused the disease. Probably the most enduring theory of eclampsia was that it was related to abdominal pressure caused by the developing fetus.[6] However, around the time of Eileen's death, three theories dominated the medical literature: the placental theory; the renal insufficiency theory; and nutrition-based theories.[7]

Advocates of the placental theory believed that toxins in the placenta were the source of the problem. According to Eardley Holland, "the placenta in the number of and power of ferments it contains is second to no organ of the body, not even excepting the liver and the pancreas."[8] Building on this observation, Holland believed that in women who developed eclampsia the level of intoxication was higher, or at least was more likely to spread outside the placenta to other parts of the system. "In light of the present knowledge," Holland wrote, "the most probable theory of the cause of eclampsia is an intoxication of the body by the passage of ferments and autolytic bodies from the placenta into the circulation."[9] But what exactly these toxins were, and why they harmed some women more than others, Holland could not say.

The placental theory developed partly in response to a series of articles documenting the strikingly similar clinical manifestations of eclampsia to various forms of poisoning, including poisoning by quinine, phosphorous, snake venom, and chloroform.[10] In one highly suggestive article published in the *British Medical Journal* in 1911, Dr. Leith Murray described in detail how cobra and rattlesnake venom affected the neurological, hematological, and renal systems, and then identified how eclampsia had very similar effects on the human system. He concluded that the anti-venoms used to treat snake bites might prove useful in treating eclampsia. "There is," Dr. Murray wrote, "considerable evidence that the pregnant woman is protecting herself against a poison directly comparable to a venom, and on this ground I wish to suggest that there may be some therapeutic use for" anti-venom in the treatment of eclampsia or the pre-eclamptic state.[11] Similarly, the parallels between chloroform poisoning and eclampsia led an American physician to recommend doctors stop using chloroform to control the convulsions associated with eclampsia.[12] Probably the strongest evidence in favor of the placental theory was a series of experiments which showed that animals injected with placental extracts died soon after the injection.[13]

Another body of research suggested that renal insufficiency was the underlying cause of eclampsia.[14] The renal insufficiency theory was based on the idea that the developing fetus introduced new toxins into the woman's system and that pregnancy could overwhelm the body's ability to purge itself of these poisons. Probably the strongest evidence in favor of the renal insufficiency theory came from studies comparing the blood and urine of pregnant women to the blood and urine of women who were not pregnant. These studies showed that while the blood of pregnant women typically had more toxins than the blood of non-pregnant women, the urine of pregnant women actually contained fewer toxins. Many physicians inferred from this surprising pattern that pregnant women were retaining more toxins than non-pregnant women, and that their renal systems were overburdened. Furthermore, there was evidence that eclamptic women were retaining more toxins than women with normal pregnancies.[15]

According to nutritional theories, the absence of specific nutrients, particularly calcium, was the primary cause of eclampsia. One of the clearest and best developed statements of the nutritional view was put forward by G. W. Theobald during the early 1930s. Theobald was puzzled by the geographic variation in the incidence of eclampsia. In London, for example, only 1 in every 10,000 pregnancies manifested eclampsia, while not far away in Glasgow, Dundee, and Kirkcaldy, 360 out of every 10,000 pregnancies manifested eclampsia. Equally puzzling was that the "incidence of eclampsia in Germany declined markedly during the [first world] war." But for Theobald the most curious pattern was to be found in Siam, southern China, and the Persian Gulf, where the "the incidence of eclampsia [was] extremely low." In these same areas, "with their teeming millions," other diseases were rampant, including "gastric ulcer, diseases of the gall bladder, [and] carcinoma of the intestines."[16]

Theobald maintained that such geographic patterns negated the placental theory of eclampsia, then the dominant theory of the disease. "I cannot believe," Theobald wrote, "that the Siamese placenta is less toxic than the Scottish placenta or that the anaemic women of Siam are more successful in providing themselves with an antitoxin than the robust women of Scotland." Theobald also wondered how the placental theory could account for the survival of roughly 50 percent of all children born to eclamptic mothers. How could a poison of sufficient strength to kill or maim the mother leave so many infants alive? "It is impossible," Theobald argued, "to explain how a diffusible [sic] toxin formed in the placenta can poison or kill the mother and allow at least half the full-term children of such women to be born alive and thrive."[17]

As for the renal insufficiency hypothesis, Theobald was no less skeptical. Claiming that eclampsia sometimes developed without inducing kidney trouble and, conversely, that kidney failure could exist without inducing eclampsia, he suggested that kidney problems were not the primary cause of eclampsia. Theobald presented evidence that changes in diet, particularly high-protein diets, could induce kidney trouble without inducing eclampsia or seizures. He also showed that exercise and physical pressure on the kidneys could cause albumin levels in urine to rise. Put another way, Theobald believed that kidney trouble might be a sign or a result of eclampsia, but it was not a primary cause of the disease.[18]

Theobold also presented direct evidence that diet might play a causal role in eclampsia. Theobold fed pregnant dogs a diet of lean meat, which provided the dogs with "all the vitamins [but] very little calcium." This diet caused "severe changes in the livers and kidneys." Moreover, the diet caused "death of the foetus if given to bitches sufficiently early in pregnancy," and Theobald hypothesized that it was "probable that human abortions, for the majority of which no adequate explanation can be offered, may be caused by deficiencies in the diet." Theobald further believed that the proteins contained in the meat, if not properly digested and absorbed directly into the human system, might have proven "extremely toxic." Based on these experiments, Theobald concluded that eclampsia was "caused by toxins absorbed from the intestinal canal, which, owing to a breakdown in the defences [sic] of the body, are not detoxicated."[19]

Eclampsia as Latent Plumbism

The doctor who first treated Eileen had his own idea about what had, and had not, caused her particular case of eclampsia. The doctor's name was Ernest E. Waters, and, according to him, the most significant aspect of Eileen's history was her prior exposure to lead. When Eileen was fifteen years old, she had started working in a lead mill. Her first attack of lead poisoning occurred just three months later, and the attacks grew increasingly severe over the next few years. The worst attack occurred when she was eighteen. For two days, Eileen was ravaged by multiple convulsive episodes; for five days, she lay unconscious; and for three weeks, she was blind. She recovered only after an eight-week convalescence. According to Waters, there was a clinically significant parallel between the convulsions Eileen experienced while employed at the lead mills to those she experienced while pregnant.[20]

Dr. Waters titled his paper about Eileen "A Case of Puerperal Eclampsia Following Lead Poisoning" and published it in the *British Medical Journal*. The most important thing about this paper is that it was wrong in at least one crucial respect. Today, the diagnosis of eclampsia is reserved for patients with no prior history of convulsions, seizures, or hypertension. Eclampsia is a pathology unique to pregnancy. Yet Eileen had developed convulsions and seizures before she became pregnant, and the diagnosis of eclampsia was incorrect. What Dr. Waters called eclampsia was almost certainly some type of latent plumbism. But while modern medicine calls the diagnostic skills of Dr. Waters into question, his more fundamental claim that Eileen's convulsions were the result of her prior lead exposure deserved wider attention and exploration. Yet with one important exception, which will be discussed shortly, it was not until the late twentieth century that researchers began to again consider the possibility that eclampsia might be related to prior lead exposure.[21]

Was Dr. Water's central argument correct? Could a woman's prior exposure to lead cause her to develop eclampsia? The answer is perhaps. As explained in chapter 2, most of the lead in the human body is stored in the bones. During the later stages of pregnancy, bone lead is mobilized, along with essential metals like calcium. If sufficient amounts of lead, from years of chronic exposure, are stored in the bone, it is possible that the mobilization of bone lead might prompt a large increase in blood lead levels and induce symptoms consistent with acute lead poisoning. More important, as the modern scientific research reviewed in chapter 2 indicates, much more bone lead is mobilized in eclamptic pregnancies than in normal ones.

The Menace and Geography of Eclampsia

In 1934, Dr. Norman Porritt published a short book entitled *The Menace and Geography of Eclampsia in England and Wales*. Largely ignored when published, the book is now long forgotten and is available only in a handful of research libraries around the world. Porritt's thesis was simple and direct. Eclampsia, he claimed, emerged with the "saturation of the [human] system by minute, infinitesimal doses of lead extending over a long period of time." The ultimate source of these "infinitesimal doses of lead" was the lead pipe used to distribute water. Although Porritt "was not so foolish as to imagine" that the removal of lead from drinking water would forever banish eclampsia, he did believe that "in its absence there would be a gratifying reduction in the number of cases and in the mortality of puerperal toxaemia and eclampsia."[22]

Porritt's evidence for this hypothesis was equally direct. Eclampsia was most common in those areas of England where the water was capable of dissolving large amounts of lead and therefore contained high lead levels. In contrast, in those areas where the water was not corrosive and individuals were exposed to little or no lead in their drinking water, eclampsia rates were much lower. For example, mortality from eclampsia was roughly 5 times greater in Halifax than in East Ham. The town of Halifax was located in the county of Yorkshire, where the water was a soft moorland variety with the power to dissolve much lead; the town of East Ham was located in the county of Essex, where the water was chalky and had little capacity to dissolve lead.[23]

The most important source of inspiration for *The Menace and Geography of Eclampia* appears to have been Porritt's own experiences as both a patient and a physician. He had practiced medicine for many years in the Yorkshire town of Huddersfield. During the 1880s, there was an epidemic of water-related lead poisoning in Huddersfield and for many years the individuals who ran the town's water company struggled with how to eliminate lead from the public water supply. Ripping up the town's lead service pipes and replacing them with pipes made of some other material was expensive and most homeowners appear to have sought cheaper modes of protection.[24] Other methods of protection included the use of paper and charcoal filters, which were effective in removing lead from tap water and were much cheaper than replacing the pipe. The town of Huddersfield also began treating the water supply with marble chips, reducing the water's power to dissolve lead. Porritt suggested that when Huddersfield began to treat the water this way there was a sharp drop in the incidence rates of eclampsia and puerperal toxemia.[25]

Over the course of his long career, Porritt had corresponded with physicians and treated pregnant women from all over England and Wales. Porritt noticed that among the female patients in areas with lead-solvent water, the rate of eclampsia and preeclampsia was unusually high, and this was true not only for the patients themselves but also for the broader population. For example, the wife of one physician was diagnosed with mild plumbism: Her urine contained 1/12 of a grain of lead per gallon, and she suffered from "flatulent dyspepsia with occasional vomiting," headaches, and appendicitis-like symptoms. She also exhibited signs of preeclampsia during pregnancy, including a "severe albuminuria." The woman lived in Westmorland, England, where the water was a soft moorland variety that leached much lead from the interior of pipes. Another case described by Porritt involved a young physician who moved to Cornwall, an area with lead-solvent water. The young physician and his wife soon developed symptoms consistent with low-grade lead poisoning, but it was not until the wife developed "puerperal albuminuria," a precursor to eclampsia, that the source of their suffering became clear.[26]

A final source of inspiration for Porritt was an article he read in the *Franco-British Medical Review*. The article was written by Dr. Leo Spira and appeared in December 1928. Spira recorded the case of an apparently healthy twenty-six-year-old woman who lived in London. Her first two pregnancies had proceeded without incident, but during her third pregnancy she developed a cold, with sneezing, cough, and headache. Within six days of delivery she became comatose, and her urine contained a small amount of albumin. Toxemia was diagnosed and a Caesarean section was eventually performed. The day after she gave birth, the patient's temperature rose to 105 degrees Fahrenheit, and she died from "paralysis of the respiratory centre." The child survived, but exhibited symptoms consistent with lead intoxication. "Two days after birth," there was "twitching of the [child's] right wrist and round the eyes and mouth. [The infant's] urine contained minute, but definite, traces of lead."[27]

Dr. Porritt on the Etiology of Eclampsia

When he wrote *The Menace and Geography of Eclampsia*, Dr. Porritt did not have access to the modern literature specifying the mechanisms through which lead exposure might predispose an expectant mother to eclampsia or preeclampsia. Consequently, his understanding of the physiological linkages between lead and eclampsia was crude, but it was not entirely incorrect. Porritt believed that lead might lie dormant in the human system for years without producing any ill effects until some shock disrupted the individual's metabolic equilibrium. "Briefly stated," he wrote, "it may be said that anything that puts a weight or an excessive load on the safety-valves of the body, adds to the possibility of latent plumbism becoming manifest." Likening the pregnant woman to a "boiler working at high pressure," Porritt argued that "the eliminatory organs"—the bowels, skin, and kidneys—acted as "safety valves." When these safety valves were "overtaxed or clogged [an] explosion" occurred.[28]

One of the more interesting aspects of Porritt's analysis of lead poisoning and eclampsia was his systematic comparison of the symptoms of the two diseases, which he claimed were strikingly similar. In terms of overt symptoms, both diseases were associated with an elevated pulse and blood pressure, kidney problems as manifested in decreased nitrogen and urea excretion, constipation, vision problems, abnormalities in reflex reactions, severe headaches, paralysis, seizures and convulsions, hearing loss, and fatigue. At the cellular level, Porritt argued that both eclampsia and lead poisoning caused changes in liver and red blood cells. Modern medical research has confirmed much of what Porritt argued with regard to cellular abnormalities among lead-poisoned and eclamptic patients.[29]

Porritt's Geography

The heart of Porritt's evidence can be reduced to a few statistics and sentences. The data in table 4.1 indicate the death rate for eclampsia and puerperal albuminuria for the period 1928–1930 in four different

Table 4.1
Eclampsia in England and Wales, 1928–1930

Region	Eclampsia-related deaths per 1,000 live births
Midlands	0.65
South	0.66
North	0.87
Wales	1.40

Sources: Porritt (1934), p. 66; Troesken (2006a).

regions in England and Wales. If lead-contaminated drinking water were an important cause of eclampsia and preeclamptic conditions such as puerperal albuminuria, one would expect those regions in England and Wales with the most lead in their water to have had the highest rates of eclampsia. According to Porritt, that is exactly what the data showed. In the north of England and in Wales, the water was soft and lead solvent, while in the south and in the Midlands, with a few important exceptions, the water was hard and had little effect on lead. Accordingly, the mortality rates from eclampsia were 32 and 112 percent higher in the north and in Wales, respectively, than they were in the Midlands and in the south.[30]

Porritt maintained that the most compelling aspect of these data was that Wales and northern England had nothing in common, except that they both had soft and lead-solvent water supplies. Wales was rural, dominated by agriculture, and had a low population density; the north of England was urbanized, dominated by industry, and had a relatively high population density: "Districts as unlike as the industrial North and rural Wales [had] this heavy mortality [and] no cause [was] common to the different areas" other than lead-solvent water. On the other hand, the Midlands and the south were as different as the north and Wales, and had only one thing in common: hard water that would not dissolve lead. There were exceptions to these general patterns, but these exceptions only helped prove the more general claim. For example, there were only a few boroughs and counties in the Midlands and the south where eclampsia rates approached those in the north and in Wales. These places included the towns of Glossop and Plymouth, and the county of Cornwall. In Glossop and Plymouth, the water was soft and lead solvent; in Cornwall, many water sources were hard but also lead solvent.[31]

Porritt's analysis of eclampsia was often placed in the context of pregnancy-related sepsis. Sepsis occurs when toxin-producing bacteria infect the bloodstream. In England and Wales during the early twentieth century, sepsis was the leading cause of maternal mortality, and it accounted for roughly one-third of all deaths in pregnancy. Deaths from sepsis outnumbered deaths from eclampsia by a factor of 3 or more, depending on the region and time period considered.[32] Porritt argued that there was very little variation in the sepsis death rate across regions and counties in England; the rate, he claimed, hovered around 2 deaths per 1,000 live births. Porritt was correct on the latter claim, but incorrect on the former. His own data suggest that there was as much geographic variation in the sepsis death rate as there was in the eclampsia death rate; the sepsis death rate was as high as 3 deaths per 1,000 live births in some counties, and as low as 0.5 in other counties.

By placing the eclampsia death rate in relation to the sepsis death rate, Porritt was able to use the death rate from sepsis as a sort of guidepost. Whenever the ratio of eclampsia deaths to sepsis deaths in a particular region was high relative to the ratios in similar areas, there was reason to suspect that the area might have had lead-solvent water and widespread lead poisoning. Using sepsis deaths this way, Porritt controlled for factors other than lead-contaminated water that might have been correlated with eclampsia. When these non-lead factors could not account for a high rate of eclampsia, Porritt suggested that the excess deaths should be attributed at least partly to lead-contaminated water.

Consider, for example, the experience of Westmorland, England. In the four years between 1919 and 1922, there were ten deaths from eclampsia and only three deaths from sepsis in Westmorland. In other words, deaths from eclampsia outnumbered deaths from sepsis by a factor of 3.3; in most situations, the exact opposite pattern obtained. Westmorland had a soft and highly corrosive water supply.[33]

In compiling his data, Porritt appears to have been systematic and thorough. He sent out questionnaires to the medical officers of towns across England and Wales, and asked them to supply figures regarding

Table 4.2
Eclampsia and sepsis in English boroughs, 1926–1930

Borough	Births	Eclampsia deaths	Eclampsia rate	Sepsis rate	Ratio
Places with lead-solvent water					
Bury	—	—	1.82	2.83	0.64
Halifax	—	—	1.37	2.51	0.55
Montgomery	2,364	3	1.26	—	—
Blackburn	7,524	8	1.06	2.13	0.50
Dewsbury	4,325	4	0.92	1.31	0.70
Swansea	13,990	13	0.92	2.29	0.40
Blackpool	—	—	0.88	2.83	0.31
Huddersfield	—	—	0.90	1.61	0.56
Westmorland	5,952	4	0.66	—	—
Places with water that was not lead-solvent					
Lambeth	9,373	4	0.42	—	—
West Bromwich	8,491	3	0.35	1.83	0.19
Eastbourne	3,418	1	0.02	1.26	0.02
West Ham	—	—	0.02	0.79	0.02
Huntingdonshire	—	0	0.00	—	—
Great Yarmouth	—	0	0.00	1.42	0.00
Worcester	—	0	0.00	0.51	0.00
Holborn	—	0	0.00	—	—

Sources: Porritt (1934), pp. 51 and 79; Troesken (2006a).
Note: Sepsis rates are for the period 1919–1922. All other data are for the period 1926–1930.

the death rates for puerperal eclampsia and sepsis for the period 1926–1930. He divided the responses into two categories, those with lead-solvent water and those without. Nine of the respondents were from places with lead-solvent water; eight of the respondents were from places with water that was not lead solvent. Table 4.2 presents a summary of his findings. The death rate from eclampsia averaged 1.09 in towns with lead-solvent water and 0.10 in towns with water that was not lead solvent. Moreover, while the death rate from sepsis was, on average, about 2 times higher in towns with lead-solvent water than in towns without, the death rate from eclampsia was 10 times higher. This can be seen by comparing the ratio of eclampsia deaths to sepsis deaths for towns with and without lead-solvent water. In the former, the ratio averaged 0.503, about 8 times greater than the average (0.045) in towns that did not use lead-solvent water.

Porritt did not believe that the other prevailing theories of eclampsia could fully account for the geographic variation in the disease. Consider Porritt's reaction to the aforementioned work of G. W. Theobald, who hypothesized that eclampsia was the result of a poor diet. Porritt was willing to concede that an inadequate diet was a "predisposing" factor for eclampsia, but questioned whether poor nutrition was the "exciting and essential [cause of] toxaemia and eclampsia." Given that many expectant mothers were underfed in England in 1930, Porritt believed that Theobald's nutritional hypothesis posited a much higher rate of eclampsia than actually existed. Moreover, while Theobald saw the geographic variation in eclampsia rates as supporting the nutrition-based theory of eclampsia, Porritt saw such variation as undermining the theory: "Toxaemia and eclampsia bear much more heavily on the mothers of the North and Wales than on those of London and the South. Are we to conclude that there is more inadequacy of diet in the North and in Wales than in London and the South?" Porritt maintained that in the north and in Wales, water was generally lead solvent, while in London and in the south, water was, with a few exceptions, hard and not corrosive of lead.[34]

Porritt emphasized that Dr. Theobald practiced in London where there were legions of poor and severely underfed people and yet London had one of the lowest rates of eclampsia in England. For the period 1928–1930, the death rate from eclampsia in London was 0.52 deaths per 1,000 births; in the north of England and in Wales, the death rates from eclampsia were 67 and 370 percent higher, respectively. According to Porritt, if nutrition-based theories were correct, one would have expected the exact opposite pattern because Londoners generally had poorer diets than those in the north and in Wales. While "nothing in the lives and habits" of Londoners could be "invoked to explain their comparative freedom" from eclampsia, there was at least one environmental factor that did distinguish London from the north and from Wales: The city's water was relatively free of corrosive agents that might have dissolved lead.[35]

Porritt was also unswayed by analysts who cited economic and social factors to explain the geographic variation in eclampsia. For example, Sir George Newman, Chief Medical Officer to the British Ministry of Health, presented data for the British counties that had the highest and lowest maternal mortality rates for the year 1931. Newman argued that in those counties and boroughs with high maternal mortality, there was a dearth of "competent [midwives,] suitable hospital accommodations, [and physicians] skilled in ante-natal supervision." Porritt countered that "the administrative counties with the lowest mortality [were] rural areas, where the measures described by Sir George Newman [were] difficult, and often impossible to obtain." Conversely, Porritt claimed that "no one [would be so] bold as to maintain that in the four county boroughs with the highest mortality, skilled aid, competent midwives, ante-natal care, and hospital accommodation [were] not within easy reach of the mothers."[36]

As for those who attributed the high rates of eclampsia in certain parts of England and Wales to industrialization and material degradation, Porritt found this unconvincing as well. He conceded that northern England was highly industrialized and included many industrial towns and cities, and he acknowledged that many of these towns had high rates of eclampsia. However, "if industrialism were at the root of the high eclampsia death rate of the Northern towns, other industrial communities, wherever situated, would [have shown] similar results." Yet "busy industrial centers" such as London, West and East Ham, Poplar, Stepney, and Deptford had much lower rates of eclampsia than non-industrial areas in the north and in Wales. Porritt pointed specifically to two villages that were not industrial but had lead-solvent water, Plymouth and Blackpool—in both places, the death rate from eclampsia was extraordinarily high. And if industrialization were the main cause of eclampsia, how could one explain the high rate of eclampsia in Wales, which was almost entirely rural and agricultural?[37]

To counter the supposition that women working outside the home was the cause of eclampsia, Porritt highlighted studies conducted by public health officials who found little correlation between female employment and eclampsia rates. Dame Janet Campbell, for example, studied eclampsia mortality in Yorkshire and she argued that, while female employment was common in the county, "the work [was] not unduly heavy and the hours and conditions [were] reasonable." Campbell believed that "the ordinary domestic work of scrubbing, washing and mangling [was] far more likely to do the woman harm." It was even possible, according to Campbell, that employment "might be beneficial." Similarly, in her study of Lancashire, Isabella Cameron "was unable to offer any explanation of the persistently high mortality in certain districts of Lancashire." Porritt explained that Cameron "could not correlate [eclampsia] with industrial employment" because there were many towns where most women went to work and yet "the maternal mortality was low."[38]

The Menace and Geography of Eclampsia, Redux

In an appendix to his book, Dr. Porritt reproduced the raw data from Campbell's "Report on Maternal Mortality." The report provided statistics on maternal mortality rates across England during the four-year period from 1919 through 1922. Although the report did not provide statistics on eclampsia, it did break down maternal deaths into two categories: deaths from sepsis (puerperal fever) and deaths from all other causes. Porritt believed that such a categorization allowed one to make rough inferences about the variation in eclampsia rates across county boroughs. Specifically, Porritt used maternal deaths from all causes other than sepsis as an approximation of the mortality rate from eclampsia. Because eclampsia was the predominant cause of maternal death other than sepsis, this was not an unreasonable approach. Sepsis accounted for one-third of all maternal deaths; eclampsia for one-sixth. Excluding deaths from sepsis, eclampsia accounted for 30 percent of all deaths. In addition, the evidence presented in the appendix suggests that eclampsia rates were orthogonal to causes of maternal death other than sepsis.[39]

Though flawed, these data can provide a starting point from which to extend the analysis given in *The Menace and Geography of Eclampsia*. For example, while Porritt often relied on small samples, Campbell's report contains information on eighty-two county boroughs in England and Wales. Analyzing this much larger sample, one finds patterns consistent with what Porritt found. The non-sepsis maternal death rate was 45 percent higher in the seventeen boroughs with a documented history of

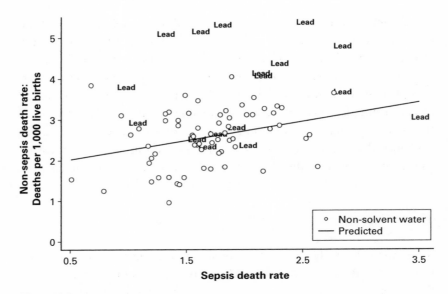

Figure 4.1
Maternal mortality and lead, 1919–1922. *Sources:* Porritt (1934), data appendix; Troesken (2006a).

lead-solvent water than in the sixty-five boroughs with non-solvent water. It is true that sepsis death rates were also higher in boroughs with solvent water, but the magnitude of that increase was smaller—the death rate from sepsis was only 20 percent higher in lead-solvent boroughs.

Figure 4.1 illustrates more clearly the relationship between non-sepsis deaths and lead-solvent water. The solid bold line is the predicted eclampsia rate for a given level of sepsis. This trend line has been estimated using only the observations for boroughs with non-lead solvent water. The observed eclampsia rates for boroughs with non-solvent water are shown by the empty circles; the observed eclampsia rates for boroughs with lead-solvent water are shown by the word "Lead." Notice that the observed eclampsia rates in boroughs with lead-solvent water, with a few exceptions, lie well above the estimated trend line. This suggests that something besides sepsis drove up the rate of eclampsia in these boroughs.[40]

There do remain many well-founded objections to Porritt's thesis and evidence. Porritt was not able to control for industrialization and urbanization. He did not have data on eclampsia rates for more than a handful of boroughs and counties, and therefore often had to rely on the maternal death rate from causes other than sepsis to approximate the eclampsia rate. This was a crude proxy. Nor did Porritt have data on overall health and well-being in different boroughs and counties; such information would have allowed him to, at least indirectly, control for factors related to diet, nutrition, and health consciousness. Finally, it was widely understood that good prenatal care reduced a woman's risk of developing preeclampsia and eclampsia. Porritt did not present any data that would have allowed him to control for variation in the quality of prenatal care across regions.

These and other problems can be addressed by going beyond the data contained in Porritt's book. Accordingly, appendix B uses data from the 1883 *Annual Report of the Register General* to explore the sources of geographic variation in eclampsia in England and Wales.[41] The results are derived from statistical models that control for potentially confounding variables (e.g., prenatal care, urbanization, and overall health) and take reasonable changes in econometric specification into account. Overall, the results corroborate Porritt's original thesis and evidence, and suggest that in regions with lead-solvent water supplies, the death rate from eclampsia was 2–3 times greater than in regions with water supplies that were not lead solvent.

Lead and Convulsions of Unknown Origin

In 1883, pregnant women were not the only people to perish after a sudden and unexpected episode of convulsions and seizure. Convulsions of unknown etiology were, depending upon the region, among the top five or ten leading causes of death in England and Wales. What caused all of these unexplained convulsions? Drawing from Dr. Porritt's insights and the modern scientific literature on lead poisoning, it is possible to suggest an answer to this question. The answer begins with a simple observation: Pregnancy is not the only physiological change that can draw lead out of the skeletal system and into the bloodstream. Animal experiments indicate that sudden changes in diet, health status, and exercise can have a similar effect.[42] Like pregnancy, these changes might alter the body's demand for metals such as calcium and iron, and can prompt the body to draw these metals out of the bones. But as with pregnancy, when the body draws out the desirable metals to fuel exercise or to help battle a disease, it also draws out undesirable metals such as lead.

This line of thought suggests the following hypothesis: Eclampsia among the pregnant and unexplained convulsions among those who were not pregnant had a common source, the actualization of latent lead stores. Three testable implications follow from this hypothesis. First, holding everything else constant, one would expect to observe a strong correlation between the eclampsia death rate and the death rate from convulsions and other nervous disorders. Second, one would expect the correlation between eclampsia and convulsions to have been driven by the presence of lead-solvent water. Third, to the extent that more bone lead was mobilized in environments where infectious diseases were common, there would have been an interaction effect between lead-solvent water and the broader disease environment. In particular, the impact of lead-solvent water would have been greater in environments where infectious diseases were common than in those where such diseases were relatively uncommon.

Appendix B tests these three propositions using data from the Registrar General's *Annual Report* for the year 1883. Although the third proposition can be rejected—there is little evidence of an interaction effect between lead-solvent water and infectious disease rates—the first two propositions survive. First, convulsions and eclampsia rates were correlated. This can be seen in figure 4.2, which plots the observed and estimated relationships between convulsions and eclampsia. The variation in eclampsia explains about 17 percent of the variation in convulsions, and for every two additional deaths from eclampsia there was one death from convulsions. Second, the statistical results suggest that, holding everything else constant, lead-solvent water increased the death rate from convulsions by 10–15 percent.

The Latent History of Eclampsia

Researchers do not normally think of studying history as a means of solving modern medical problems. The reason for this is both simple and logical. As an empirical science, medicine is predicated on testing

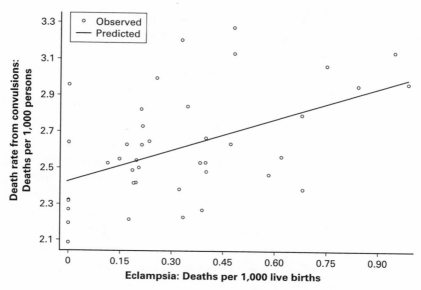

Figure 4.2
The correlation between eclampsia and convulsions, 1883. *Source:* Registrar General (1883). *Note:* The R^2 on the trend line is 0.173, with an estimated slope of 0.492, which is significant at the 0.003 level (one-tailed test). The regression and graph exclude five outlying districts ($N = 42$). See appendix B for a more complete statistical model.

and validation: Theories about specific diseases are confronted with data and those that fail to survive the confrontation are invalidated and replaced by theories that better explain the data. Consequently, the present state of medical knowledge embodies all that one needs to know about history: Those theories that have proven correct, regardless of when they were first developed, are already known to current medical practitioners; the only historical theories that are not known to current practitioners are the ones that have been proven wrong. Within this construct, the study of history is little more than the study of mistaken ideas, and it is not immediately obvious why anyone would want to invest much time studying the intellectual mistakes of previous generations.

If improvements in life expectancy and eradication of disease are the benchmarks, it is clear that the ahistorical approach to medicine has been successful. It is, however, possible to argue that a more historical approach might have some unappreciated benefits. There are at least four reasons to think that this is so. First, theories can reveal important insights and truths about the world even if they are, in some fundamental respect, incorrect. Second, historical patterns in disease rates often reveal important clues about the causes of particular diseases, and about how to best prevent them. Third, when an early generation of scientists ignores, for whatever reason, an unorthodox and original thinker in their field, subsequent researchers will not be aware of the ideas promulgated by that thinker, even if those ideas are largely correct and contain important insights regarding disease prevention and treatment. Finally, there is no question that earlier generations of scientists had to rely on relatively crude technologies and experimental techniques when they tested hypotheses. To the extent that this more primitive science led researchers to mistakenly and prematurely reject theories and hypotheses that were true or at least partly true, it might be beneficial to current researchers to reconsider historical theories and evidence.

When any of the four conditions specified above is satisfied one might describe the relevant history as latent: The past lies unseen and un-exploited, but it is nonetheless ripe for scientific inquiry. The history of eclampsia is latent. It satisfies at least three of the criteria specified in the foregoing paragraph. In particular, although the early research on eclampsia was full of mistakes, and is now ignored partly because of those mistakes, that research contains useful insights. Similarly, historical patterns in eclampsia rates suggest a strong correlation between the chemical characteristics of public water supplies and the incidence of eclampsia. But the most significant source of latency was that doctors around 1900 chose to ignore or dismiss research showing that prior lead exposure greatly increased a woman's risk of developing eclampsia. Only in the past ten years have scientists been reawakened to the possibility that high lead levels might increase the risk of eclampsia.

Summary

This chapter has reviewed a now long-forgotten work, Norman Porritt's *The Menace and Geography of Eclampsia*. In this book, Porritt argued that eclampsia rates in England and Wales were correlated with the lead levels in local water supplies. The statistical evidence reviewed here supports Porritt's original contention. In places with high water-lead levels, eclampsia rates were 2–3 times greater than in areas with low water-lead levels. These patterns are consistent with the scientific evidence reviewed in chapter 2 showing that more lead is mobilized from the maternal skeleton in eclamptic pregnancies than in normal ones. Also, the effects of lead-contaminated water were not limited to convulsions during pregnancy, but increased deaths from convulsions in general.

5

The Secret of Dr. Porritt's Society

Around 1910, Norman Porritt himself had been afflicted with a strange malady that eluded diagnosis, both his own and that of the other physicians he consulted. Dr. Porritt was so baffled and intrigued by his affliction that he published a lengthy article about it in the *British Medical Journal*. According to Porritt, a "strange lethargy" crept over him. "Feeling a weariness of flesh and brain," he gradually lost interest in life and withdrew from family friends. At social gatherings, he refrained from joining in conversations and would sit "staring at the fire" overcome by gloom. His bowels were "constipated and stubborn," and he derived "no satisfaction from food." Sleep was his only escape from his increasingly meaningless existence, and for Porritt, the desire to sleep was overwhelming. The doctor had to muster all of his strength to perform even the most mundane tasks: "dressing himself," "bathing," and "reading" were all ordeals "he wished he could shirk." Absent any identifiable physical pathology, the doctor's family and friends attributed Porritt's new-found melancholy to a character flaw, and as his melancholy grew, so too did their contempt for him. This, in turn, further isolated Porritt and compounded his sadness.[1]

At one point, Porritt thought it might be his thyroid and so he "dosed himself" with a thyroid medication. But it did not work. A fellow doctor recommended that he get "a change of air," and so Porritt removed himself from his work and home for a long vacation. After several weeks away, he felt better, but upon returning home, the oppressive lethargy returned. Eventually, Porritt had his urine and domestic water supply tested. His urine contained 8/19 of a grain of lead per gallon, and the domestic water supply contained 1/14 of a grain per gallon (about 80 times the modern EPA standard). With this discovery, Porritt stopped drinking water from his household tap, and he soon regained his vitality. All was fine until some years later, when he moved. Once again, a strange lethargy crept over him, and after some time, the doctor felt estranged and oppressed by life's smallest tasks. Weary and confused, Porritt did not immediately recognize that his old illness had returned, but once he did, he had his urine tested for lead. It contained 4/25 of a grain per gallon. Upon finding a new water supply, his health was restored.[2]

After recovering from his afflictions, Porritt met a young doctor who had developed a similar illness. Like Porritt, this doctor had been slowly overtaken by weariness and lethargy, which he attributed to an "enervating climate." Porritt recommended that his young colleague have his urine and tap water tested for lead. The doctor "scoffed" at the suggestion. A short time later, Porritt heard that this same doctor's wife "had been laid aside by" kidney problems associated with childbirth. Porritt had long believed that lead exposure increased the risk of maternal mortality during childbirth, so he approached his colleague and again urged him to have his urine and tap water tested for lead. This time, his colleague relented. The subsequent chemical analysis revealed high lead levels in the young doctor's urine and tap water. Three weeks later Porritt received a letter from his colleague stating that the lead had been eliminated from his tap water and that he "felt much better" and was "not so lazy." Six weeks later the young doctor pronounced himself "free from any of the symptoms of plumbism."[3]

With this experience, Porritt's young colleague soon became "alive" to the possibility that lead-tainted water might be the source of many unexplained and puzzling afflictions. The doctor had eighteen of his patients' urine tested for lead. In eight of the patients no lead was found, but in the remaining ten, lead levels in the urine ranged from 1/125 to 1/5 of a grain of lead per gallon. Those patients with the highest lead levels improved after steps were taken to eliminate lead from their drinking water.[4]

Porritt himself believed that there were two types of plumbism: "One, the classical lead poisoning of the textbooks, caused by massive doses," and "the other," a "slow, subtle insidious saturation of the system by infinitesimal doses of lead extending over a long period of time, and producing a group of symptoms altogether different from the recognized forms of plumbism." In the latter form of plumbism, the typical tell-tale signs of lead poisoning were rarely observed: colic was "rare"; a blue line seldom appeared on the gums; and wrist-drop occurred "only if the significance of the early symptoms" was "not recognized." Porritt also claimed that anemia was "rarely noticeable."[5]

Was Porritt's self-diagnosis correct? Did he really suffer from low-grade lead poisoning brought on by contaminated drinking water? Several pieces of evidence suggest the answer to both questions is yes. First, both Porritt's drinking water and urine contained nontrivial amounts of lead. Second, Porritt's symptoms improved after he stopped drinking water from his home, and they returned once he resumed drinking water. Third, Porritt's symptoms are consistent with those one might observe in individuals chronically exposed to low doses of lead.[6] Finally, the case study recounted by Porritt is almost identical to a case study recently published in the *Lancet*—the latter, a case that was, without question, lead poisoning.

In the recent case, a forty-seven-year-old woman was admitted to an Australian hospital in 1995 or early 1996. Much like Porritt, the patient suffered from general debility, headaches, confusion, weight loss, constipation, and "had difficulty with abstract reasoning." Tests revealed a blood-lead level 9.75 times the level considered safe by Australian officials at the time. Further study revealed that the sole source of lead exposure in this case was lead-contaminated water. In particular, lead-based solder was used in the patient's hot water heater, and she was a heavy coffee drinker. Ten years prior to her admission to the hospital, the patient's symptoms had been attributed to chronic fatigue syndrome, and for ten years, that is what the patient and her physicians believed. In fact her symptoms had been caused by drinking lead-contaminated water.[7]

Ad Nauseam

Dr. Porritt's account of this "slow and subtle" form of lead poisoning is remarkable. Rarely does one find a doctor so willing and able to empathize with his patient, perhaps because in this case they were one and the same. But more important is Porritt's central message: Diagnosing water-related lead poisoning was difficult. Porritt was an accomplished physician and surgeon who wrote numerous articles and seven books over the course of his career, yet he was very slow to diagnose his own bouts with water-related lead poisoning.[8] Even when he developed the disease a second time, an accurate diagnosis eluded him. This was a disease that went unnoticed or inaccurately diagnosed by all but the most sensitive and observant physicians. And for those physicians who believed that water-related lead poisoning was only a remote possibility, it is highly unlikely that they would have ever observed such cases in their own practices—not because such cases had never passed through their office doors, but because when they did, the doctors looked in another direction.

Porritt's experiences with water-related lead poisoning were not unusual. The medical literature was, and still is, full of similar stories and case-studies. In many of these studies, individuals would not be properly diagnosed for years and, in some cases, so much lead accumulated in their systems that they died of lead poisoning before anyone was able to identify the true cause of their suffering. Consider one of the first cases of water-related lead poisoning documented in the English literature. The events occurred in Worcester, Massachusetts, during the early 1700s. A certain husband and wife had twenty one children: eight of these died young; the other thirteen outlived their parents. "During their infancy, and indeed until they had quitted the place of their usual residence, they were all remarkably unhealthy; being particularly subject to disorders of the stomach and bowels." The parents too had very poor health: "The father, during many years, was paralytic; the mother, for as long a time, subject to colics and bilious obstructions. She died at last of an obstinate jaundice."[9]

While the parents were alive, accurate diagnosis eluded every physician who treated the family. Many times the family traveled far away for treatment in mineral waters, and with this therapy the disease subsided, only to reappear soon after they returned to their home in Worcester. Only when the father died and the children were forced to sell the family home was the true cause of their lifetime of suffering revealed. The buyers of the home discovered the water pump and associated piping, which were made of lead, needed serious repair: "The pump was found to be so corroded, that several perforations were observed in the cylinder" and "the cistern in the upper part was reduced to the thinness of common brown paper, and was full of holes, like a sieve."[10]

A plumber who had worked at the family home later said "he had repaired the pump several times" over the previous years. On each of the occasions he found the water pump corroded. The plumber was convinced that the corrosion was "effected in a short time" and that the water "must have been very strongly impregnated with the noxious qualities of the metal." A doctor who had long cared for the family subsequently wrote that "the foregoing account fully confirms...that the water of this pump mixed with lead, did occasion the unhealthiness of the family, who drank it." But it was not until eight children and two adults died, very likely from lead-related causes, that the true source of their suffering became apparent.[11]

In 1839, James Alderson reported two instances of what he described as "cases of paralysis from the unsuspected absorption of lead, in consequence of drinking rain water, kept in lead cisterns." These cases were discovered when another physician requested Alderson's assistance with a patient whose ailments baffled him. That patient, a Mr. Thackery, was a sixty-three-year-old man who had long been "laboring under paralysis of the upper extremities, and partial paralysis of the lower." He had limited power in his arms and hands, and to move to and from his bedroom, "he required the assistance of a servant on each side of him, and then his knees bent under him, and his gait was tottering." Even with the aid of a stimulant, the patient's bowels acted only once every three or four days, causing much abdominal pain and distress. Much like Dr. Porritt, the patient also suffered from melancholy and "would frequently shed tears from light causes." At one point, Thackery had a seizure in which he fell out of bed and dislocated his shoulder.[12]

The source of Thackery's suffering escaped his physicians until it was recognized that Thackery's sister-in-law, who had lived with him, had the previous year been attacked by paralytic symptoms identical to those of Thackery. The sister-in-law eventually died "without any cause of the paralytic symptoms having been made out." This struck Alderson as a "remarkable coincidence" and prompted him to examine the household's water supply as a possible source of lead poisoning. He discovered that the household used rain water, "collected from the top of the house by means of lead gutters, and received into a cistern lined with lead." The water had a "sweetish taste" which the servants in the home had "often remarked" upon. Alderson consulted a chemist who found evidence that the water was eroding the inside of the leaded cistern. Once Mr. Thackery stopped using this water, he regained his ability to walk and partially regained the power in his wrists and fingers. It took several years and one death, however, before physicians were able to isolate the cause of the suffering in the Thackery home.[13]

Writing in *The New Edinburgh Philosophical Journal* in 1859, Dr. Lauder A. Lindsay hypothesized that water-related lead poisoning was common but rarely diagnosed. Lindsay "firmly believe[d] [that] cases of lead poisoning [were] constantly occurring in all [the] large towns from the plumbeous [sic] impregnation of drinking waters." Physicians were often "extremely puzzled [by the] anomalous symptoms" associated with water plumbism, Lindsay wrote. Citing the experiences of many of his colleagues, he argued that "many obscure cases of colic and other intestinal affections [sic], as well as of paralysis of the nature of lead palsy—sometimes going on to a fatal issue [were] really due to plumbeous [sic] impregnation of drinking-waters."[14]

William Thomson, a chemist, reported having been summoned to a home in England in 1879, where he found the water in the home contained 0.197 grains of lead to the gallon, 225 times the modern EPA standard. A large family lived in the house. Most of the occupants enjoyed only "indifferent" health, but one individual, a woman, suffered from a serious case of lead poisoning. The gums of the woman in question "were tinged of a bluish shade and the fingers of both hands had become stiff and partially paralyzed." After the source of this suffering was discovered, the home's lead water pipe was removed. The woman eventually recovered and even those who were not so severely afflicted exhibited improved health "after the removal of the lead pipe." According to Thomson the unusual features of this case were twofold. First, the afflicted woman diagnosed herself, while the physician who was treating her had been puzzled for years. Second, the family lived in the home for twenty-one years before discovering the true cause of their malaise. After describing this case, Thomson speculated that it seemed "probable that many persons may be suffering from slow lead poisoning without the real nature of the malady being recognised by medical men."[15]

Another example of water plumbism failing to be diagnosed in a timely manner took place in a convent near Lyons, France. For ten years, the nuns in the convent consumed water containing 2.3 milligrams of lead to the liter, 153 times the modern EPA standard. A French physician discovered the case only after three nuns had died of lead poisoning and the surviving twelve nuns had become "seriously affected."[16]

While cases such as these could be repeated it might be useful at this point to summarize the views of experts on lead poisoning in general, and water-related lead poisoning in particular. Writing in 1914, Sir Thomas Oliver stated that "lead poisoning arising from water supply" often gave rise to symptoms which were "not typical of those observed in occupational cases" and, unless there was "some other reason to suspect plumbism" it was "quite easy to overlook their true cause." After studying an epidemic of lead poisoning in Diepholz that was caused by impure water, Helwes considered "the changing manifestations" of the disease "very puzzling" and claimed that "water plumbism" was easily "overlooked in diagnosis" as a result. In a doctoral dissertation published in 1897, Ebner of Wüzburg offered similar sentiments.

In one widely discussed outbreak of water plumbism, a French physician to the royal family confessed that he had been "very puzzled by the abnormal pains suffered and the complex symptoms presented" when a large contingent of the royal family was made sick by water lead. After an outbreak of plumbism in Sheffield, England, during the 1880s, the local health officer, Sinclair White, wrote that even though local physicians were "quite familiar" with occupational forms of lead poisoning—there were many lead mills in the city—those same physicians had great difficulty diagnosing water plumbism and often could not do so until public-health officials alerted them to the possibility.[17]

Even today, with the aid of late-twentieth-century technology and medical training, lead poisoning can be a difficult and elusive diagnosis. Consider two recent cases. The first case involved a two-year-old girl in New Hampshire. The case was reported by the Centers for Disease Control and Prevention in 2001. The child was admitted to the emergency room of a community hospital with vomiting and a low-grade fever. A throat swab for a streptococcal infection was positive, and the girl was discharged with a prescription for an appropriate antibiotic.[18]

The girl's vomiting continued, however, and she was re-admitted to the same hospital two weeks later. The next day she was transferred to a tertiary-care hospital. Within a few hours of her transfer, she became unresponsive, and had difficulty breathing. Doctors placed her on a respirator and performed a brain scan. The brain scan revealed diffuse cerebral swelling and dilated ventricles. A blood test later showed that the girl had a blood-lead level of 391 μg/dl; any blood lead level above 100 is potentially fatal. It was subsequently observed that in the apartment where the girl had lived, there was peeling lead paint. Chelation therapy was administered, but it was too late. The girl became comatose and died two days later.[19]

The second case involved a two-year-old girl in a small village just outside Bangkok. The girl's symptoms first appeared in the early summer of 1976. Her hands became weak and she had trouble picking things up. Soon she developed a high fever and began passing out. Her parents took her to a navy hospital near the village, where the doctors diagnosed her with some sort of an infection and prescribed an antibiotic. But the treatment did not work and, in June 1976, the girl developed severe diarrhea and began throwing up uncontrollably. Her parents then took her to Chulalongkorn University Hospital in Bangkok, where doctors were still unable to accurately diagnose her illness. Three days after she was admitted to the hospital, the girl died, perishing in a fit of convulsions.[20]

According to medical observers at the hospital in Bangkok, the girl's death would have escaped notice had it not been for a young intern who thought the death was suspicious. An autopsy was ordered and only then was it discovered that the girl had been lead poisoned. Eventually it was learned that forty-nine children in the village where this girl had lived were lead poisoned; in twenty of these cases, the exposed child had an "extreme" lead concentration. Soon after the death of her daughter, the girl's mother realized that she had "recently felt dizzy quite often." The sources of exposure in the village were traced back to a road that had been paved with old lead batteries, the town's food supply, and the town's drinking water.[21]

A State-of-the-Art Diagnosis

During the 1800s, physicians mainly depended on two types of information to diagnose lead poisoning: lead levels in urine and the observed symptoms of the patient. Urine tests enjoyed the support of many prominent physicians, including Sir Thomas Oliver. According to Oliver, lead in the urine occupied "the same relationship to saturnism [as] Koch's bacillus [did to] tuberculosis...Eberth's bacillus to typhoid fever...and ...Klebs-Loeffler's bacillus to diphtheria." In other words, if one observed lead in the urine, there was a high probability that the individual had, or would develop in the near future, lead-related afflictions. While Oliver might have been correct on this point, he and others who relied on the urine test often implicitly assumed that the corollary was also true—that is, if lead was not observed in the urine, the individual was not lead poisoned. This was not always the case. An individual could have been seriously lead poisoned and yet eliminate little or no lead through the urine. In a study of lead poisoning among painters that was published in 1915, more than half of the 162 cases of active lead poisoning had no lead in their urine. Analysis of urine was a poor indicator of lead poisoning because it measured the amount of lead the body was excreting, as opposed to the amount it retained, which was what determined poisoning.[22] Doctors who relied heavily on urine tests to diagnose lead poisoning would have mistakenly diagnosed many victims of lead with some other ailment or affliction.

The unreliability of urine tests was largely moot, however, because most nineteenth-century doctors appear to have relied mainly on the patient's symptoms to diagnose lead poisoning. Most doctors sought to identify lead poisoning by examining their patients for a symptom, or set of symptoms, that satisfied two criteria: first, the symptom was experienced by all, or nearly all, victims of lead poisoning; and, second, the symptom was unique to lead poisoning. Of lead's many symptoms, the ones that came closest to satisfying these criteria were: colic; paralysis in the extremities, particularly wrist- and foot-drop; and the blue gum line. Unfortunately, trying to diagnose lead poisoning through these symptoms was probably not any more effective or reliable than using a urine test.[23]

As described by one British doctor, lead colic typically manifested itself as a "sharp stabbing pain" in the abdomen or as an "acute twisting" pain that felt as if "the bowels were being nipped." Attacks of lead colic usually "came on quite suddenly and with such severity that the patient would roll about in considerable agony." The duration of the attacks varied from "a few minutes to several hours." Colic was a useful diagnostic tool because it appears to have been observed in the majority of cases of lead poisoning. An article by Stainthorpe published in the *British Medical Journal* in 1914 described the symptoms attending 120 cases of water-related lead poisoning. In 102 (85 percent) of these cases, doctors identified some form of abdominal distress or colic.[24] A much earlier and larger study, conducted during the early 1800s by a French physician, found that of 1,493 cases of lead poisoning, 1,207 (81 percent) suffered from colic.[25] One problem with using colic as indicator of lead poisoning, however, was that it was not unique to that sickness, and similar pains could have been caused by such things as appendicitis and gall bladder dysfunction. There were published reports of lead-poisoned individuals having had their appendix or gall bladder surgically removed only to have the true cause of their suffering discovered later.[26]

The blue gum line was the most widely used indicator of lead poisoning. As described by Tanqueral in the 1840s, "the first and most frequent symptom of the presence of lead in the system is a very peculiar discoloration of the gums and teeth."[27] Another authoritative source published some eighty years later concurred, describing the lead line as the "most constant sign of plumbism" and one that appeared in almost "no other condition."[28] Evidence from these and other sources indicates that many, perhaps most, doctors refused to diagnose a patient as lead poisoned unless the blue gum line appeared, no matter how indicative the patient's other symptoms were of lead poisoning.[29]

While the blue gum line was an unusual symptom that was almost always caused by lead exposure (although phosphorous poisoning could also cause a blue gum line), relying solely on the gum line as an indicator of lead poisoning was a mistake. The blue gum line appeared in only a minority of all cases. In the aforementioned study of lead-poisoned painters, it was found that less than 14 percent of the patients exhibited a blue gum line.[30] Similarly, in the aforementioned study by Stainthorpe, it was found that of 120 victims of water-related lead poisoning, only fifteen (12.5 percent) had a blue gum line.[31] For infants the blue gum line was even more rare: Of 298 lead-poisoned infants treated at the Kyoto University Hospital between 1902 and 1923, only 1 exhibited a blue gum line, despite the fact these were severe cases with a case mortality rate of 59 percent.[32]

Like the gum line, wrist- and foot-drop were symptoms not easily attributable to causes other than lead. In wrist-drop, the hand hangs at a sharp right angle to the arm and the fingers curl slightly underneath the palm and wrist. There were, however, two problems with using paralysis and wrist- and foot-drop as indicators of lead poisoning. First, these symptoms probably occurred in only 10 percent of all documented cases of lead poisoning.[33] Second, these were extreme symptoms. If an individual ingested enough lead to lose motion in his or her hands, arms, or feet, that individual would have already ingested significant amounts of lead and done serious and perhaps irreparable harm to the body.

That symptoms like paralysis and wrist-drop were too crude a diagnostic tool had become clear to at least some doctors by the early 1900s. One of the most authoritative statements on the diagnosis of lead poisoning was written by Harry Linenthal and published in the *Journal of the American Medical Association* in 1914. Linenthal was a physician at the Massachusetts General Hospital in Boston, and the State Inspector of Health for Massachusetts. According to Linenthal, there was a "tendency" among doctors to delay the diagnosis of lead poisoning until objective signs "pathognomic to the disease" were observed. "This tendency [was] dangerous [and caused] incalculable harm." Likening lead poisoning to tuberculosis, Linenthal wrote, "The physician who fails to recognize tuberculosis [without] the presence of tubercle bacilli in the sputum very often fails to avail himself of the opportunity of arresting the disease in its incipiency." The same was true of lead poisoning. "Waiting for certain signs before establishing the diagnosis [likely allows] the poison to undermine the constitution [beyond any] remedial measures."[34] One might quibble with Linenthal's comparison of lead poisoning and tuberculosis. A better comparison could probably have been made between lead poisoning and diabetes. By the time a victim noticed the symptoms of diabetes—say blurred vision—he or she might have already damaged the heart and kidneys. But Linenthal's larger point remains valid.

It is instructive to place Linenthal's diagnostic analysis in context. Suppose physicians and public-health officials used as their barometer of unsafe lead levels in water the answer to the following question: Did officials observe any diagnosed cases of lead poisoning in the community traceable to the public water supply? If not, then lead levels were presumed safe. If so, then lead levels were unsafe. This heuristic is no historical contrivance. As will be made clear in subsequent chapters, people quite rightly demanded proof before they were willing to believe that lead water pipes were dangerous, and the proof they demanded was in the form of documented cases of water-related lead poisoning. (See, e.g., the discussion of Glasgow in chapter 8.) But if public-health officials waited until people were becoming paralytic because of the water, the health effects on the broader population probably would have been quite severe.

More important, if there were sufficient lead in the water to paralyze an adult, it is not difficult to imagine how that same water would have affected a developing fetus or infant. Indeed, animal studies have shown that the amount of lead needed to induce fetal or infant death is far below the level needed to induce the clinical symptoms of lead poisoning in developed animals.[35] A pregnant mother probably could have imbibed lead-contaminated water without developing any outward symptoms of lead poisoning, while at the same time exposing her developing fetus to a fatal dose of lead. Consider the case of C.B., a thirty-one-year-old housewife who lived in Enfield, Massachusetts, during the 1920s. For four years, C.B. had been consuming tap water that contained 2.261 milligrams of lead per liter, and her daily intake of lead was 6.214 milligrams. The lead levels in C.B.'s tap water exceeded the modern EPA guideline by a factor of 150, and her daily intake of lead exceeded the daily dose of lead contained in leaded abortion pills by a factor of 24 (see chapter 3). Yet C.B.'s own physical symptoms were mild and included only abdominal pain and constipation. Although she had one child, it is not clear if the surviving child was born before or after C.B. began consuming the leaded tap water. Moreover, C.B. had a history of "early pregnancy." C.B.'s husband, who consumed the same water, exhibited no overt symptoms of lead poisoning.[36]

The fallibility of diagnoses based solely on symptoms prompted Linenthal to advocate an alternative mode of identification. According to Linenthal, the "all-important factor" in diagnosing lead poisoning was the "history of exposure."[37] If an individual worked or lived in an environment with a high degree of lead exposure, it was possible that, whatever the patient's particular affliction or symptoms, the sickness was caused by lead. Hence, from Linenthal's perspective, once it was established that a patient had been exposed to lead, the practitioner confronted one question: Was the level of lead exposure sufficient to induce the symptoms observed in the patient? If not, then other potential illnesses needed to be considered. If yes, then the physician should proceed to administer therapies designed to reduce lead in the system, and to proscribe any further lead exposure for the patient. Given the wide variety of symptoms that lead induced (see chapter 2), this was a reasonable way to proceed and would have been highly effective as long as physicians were sufficiently sensitive to the effects of even low doses of lead.

Once again it is useful to place Linenthal's analysis explicitly in historical context. Many nineteenth-century doctors simply did not believe that enough lead could be leached from the interior of water pipes to cause serious health effects.[38] If a doctor's training and analytical framework led him to believe that water-related lead exposure was minimal and perfectly safe, he would have looked elsewhere for the source of his patient's illness. As one British physician remarked in discussing the tendency to mistakenly attribute genuine cases of lead poisoning to other causes: "More mistakes are made by not expecting than by not knowing."[39]

By the mid-twentieth century, the techniques for diagnosing lead poisoning had much improved, as physicians began examining hair and blood for the presence of lead, and innovations in urine testing made those tests more reliable. According to C. N. Myers and multiple coauthors, improved testing revealed "a much higher rate of lead poisoning" than had "previously been suspected." Echoing Dr. Porritt's diagnosis, the Myers study concluded that "the cumulative effects of small doses of lead" were the cause of "many obscure symptoms" that had long been "missed by physicians." Myers believed that water-related lead poisoning was particularly easy to miss because at low-grade exposure lead poisoning often mimicked rheumatism.[40]

The Infectious Disease Environment

The disease profile in nineteenth-century cities was dominated by infectious and epidemic diseases that affected the young disproportionately. Before 1880, probably more than half of all deaths in large American cities occurred among children less than five years old, and the leading causes of death were diarrheal diseases—such as typhoid fever, cholera infantum, and dysentery—and respiratory diseases—such as tuberculosis, influenza, bronchitis, and pneumonia. In contrast to the lead poisoning observed in Norman Porritt, these diseases typically killed swiftly and in unmistakable ways.[41]

For example, once ingested, the cholera bacillus multiplied rapidly in the alimentary tract, producing "violent and dramatic symptoms." The victim experienced "massive vomiting and diarrhea" and lost as much as one quarter of his or her bodily fluids and "essential salts." Within a few hours' time, the patient was reduced to a "comatose, apathetic state, with sunken eyes and blue-grey skin." Roughly one-half of all cholera victims died, and they often perished less than twelve hours after their first symptoms appeared. "Because death came so quickly, "perfectly healthy people," whatever their socioeconomic status, never felt safe "when the infection was anywhere near." The fear surrounding cholera was compounded by its non-Western origins. Endemic in parts of Asia, cholera was unknown to Europeans until 1832, when it spread rapidly from east to west—out of Asia and across Russia, to continental Europe and Great Britain, and finally to North America.[42]

Yellow fever was equally frightening. Spread by mosquitos, yellow fever would lie dormant for years and then erupt suddenly in large port cities, particularly those in the American South. As its name implies, yellow fever adversely affected liver function (resulting in jaundice) and caused a high fever. Other symptoms included headache, restlessness, chills, and nausea. For those who survived, the disease reached its peak three or four days after the onset of symptoms. For those who did not, the disease eventually caused kidney failure and internal hemorrhaging that manifested itself in blackened vomit and bleeding from the nose and mouth.[43]

Deadly and highly contagious diseases naturally took precedence over lead poisoning. Nobody in the nineteenth century doubted the capacity of cholera, yellow fever, and typhoid fever to kill thousands of people within a few months' time. There were, however, a great many people who doubted the capacity of low-grade lead poisoning to induce anything more than the occasional case of colic, rheumatism, or fatigue. As explained in chapters 2 and 3, the hypothesis that low-grade lead exposure increased the incidence of miscarriages, stillbirths, and infant mortality developed slowly. While a handful of physicians made such arguments in the 1880s and 1890s, it took more than a century for medical researchers to accumulate the evidence necessary to really sustain such a claim.

Faced with the choice to combat diseases that were, without question, killing thousands of people every year, or a disease that might have been inducing the odd case of colic, most physicians and public health officials chose to focus on the former. For scientists, the choice would have been less research dedicated to lead and its effects, and more research on typhoid, cholera, and the like. For the practicing physician, the result would have been triage, giving patients with the most pressing ailments first priority.

On the Possibility of Silent Epidemics

How frequently did nineteenth- and early-twentieth-century doctors mistakenly attribute water-related lead poisoning to some other illness or ailment? Although it is impossible to answer this question with precision, there does exist some evidence. Dr. James J. Putnam, an instructor at Harvard Medical School, surveyed inpatients, outpatients, and medical students at hospitals in the Boston area, particularly the Good Samaritan Hospital and Massachusetts General Hospital. None of the subjects had been diagnosed as lead poisoned; they were receiving treatment for ailments or illnesses that were not thought to have had a basis in prior lead exposure.[44]

There were 186 subjects in the study. Dr. Putnam collected urine specimens for each subject and analyzed it for lead content. He found undue lead levels in 40 percent of those examined. Putnam inferred from this that lead was likely "a partial cause of a variety of symptoms not usually attributed to that cause."[45] Another possible conclusion is that lead exposure was much more widespread than Dr. Putnam and his colleagues realized. Subsequent research would bear out the latter, but the discovery that lead exposure was pervasive even among otherwise apparently healthy individuals was unfortunately interpreted as evidence that it was "normal" for healthy human beings to have lead in their system. That is, a little lead in the system did no harm. Not until the past few decades has the so-called normalization of lead come to be abandoned.[46]

Where did Dr. Putnam believe all of the lead exposure in Boston came from? Although he offered no precise calculations on the matter, he concluded his essay by highlighting cases of water plumbism in the city. He also cited recent studies showing that Boston tap water often contained lead. To deflect the argument that Boston water did not contain sufficient levels of lead to seriously impair health, Putnam asserted that there was no known threshold of what constituted a safe exposure level. "It is never safe," the doctor wrote, "to say how large a quantity of lead a person may bear, nor how small a quantity may effect [sic] him badly." Furthermore, Putnam suggested that lead poisoning, particularly water-related lead poisoning, was underdiagnosed on a wide scale.[47]

The most compelling evidence that there could be widespread and unrecognized epidemics of water plumbism comes from a study conducted by Wade Wright, Clarence O. Sappington, and Eleanor Rantoul. Wright and Sappington were physicians affiliated with the Massachusetts Department of Public Health who in the summer of 1923 surveyed 253 persons from twenty-seven cities and towns in eastern and central Massachusetts. They gathered data on each person's water supply and the length of lead service pipes to the person's home. They also gathered medical information that could be indicative of lead poisoning, including blood smears, a hemoglobin test, and examination of gums for a lead line. In conducting the blood tests, Wright and Sappington were looking for evidence of stippling (abnormalities in the size and shape of red blood cells), an indicator of lead poisoning. Their goal was to estimate the extent of water-related lead poisoning in these twenty-seven towns, and to draw inferences about the incidence of such poisoning in the broader New England region.[48]

Wright and Sappington designed their sample so that it would be representative of the population. Cities of all sizes—"large cities, moderate-sized towns, and small villages"—were included in the sample, and the "economic status of those persons observed varied greatly." Significantly, the sample was not organized around surveying known cases of lead poisoning. Nor does it appear that any of the sample respondents had ever been diagnosed as lead poisoned by physicians other than Wright and Sappington.[49]

Collaborating with Eleanor Rantoul, a statistician with the Metropolitan Life Insurance Company, Wright and Sappington published their findings in an article in the *Journal of Industrial Hygiene*. Of the 253 persons surveyed, 63 (25 percent) were diagnosed as lead poisoned, and in each case the exposure was traced back to the individual's water. Moreover, the authors adopted a very strict definition of lead poisoning: "In addition to a lead line or stippling of the red blood cells, at least two symptoms common to cases of lead poisoning" had to have been present before the individual was diagnosed as lead poisoned. In light of this definition, 20 individuals with stippled red blood cells and 2 individuals exhibiting a blue gum were excluded from the poisoned category because they exhibited no other overt symptoms of plumbism. If these 22 individuals had been diagnosed as poisoned, 33 percent of the sample would have been designated as lead poisoned.[50]

To the extent that the Wright and Sappington sample was representative of New England populations, it suggests that between one-quarter and one-third of the population that employed lead piping was lead poisoned due to lead-contaminated water. In Massachusetts, 39 percent of the population used lead water pipes (see appendix A, table A.1, weighted sample), suggesting that between 10 and 12 percent of the state's population suffered from water plumbism.[51]

Table 5.1 provides a summary of the principal symptoms found among the 253 persons examined. Two conclusions emerge from the table. First, most of the symptoms observed suggested mild to moderate

Table 5.1
The incidence of water plumbism in Massachusetts

Finding	Total cases		Poisoning cases	
	Number	%	Number	%
Total cases	253	100.0	63	100.0
Pallor	157	62.1	47	47.6
Low hemoglobin	82	32.4	25	39.7
Constipation	78	30.8	25	39.7
Eructations	77	30.4	26	41.3
Stippling	76	30.0	56	88.9
Headache	73	28.9	33	52.4
Joint pain	57	22.5	14	22.2
Abdominal pain	53	20.9	22	34.9
Vertigo	36	14.2	9	14.3
Lead line	26	10.3	24	38.1
Weakness in forearm	26	10.3	14	22.2
Loss of appetite	21	08.3	9	14.2
Weight loss	15	05.9	8	12.7

Source: Wright, Sappington, and Rantoul (1928).

lead poisoning, and it is easy to see how these symptoms could have been overlooked, trivialized, or improperly diagnosed by physicians unattuned to the dangers of lead. The primary symptoms included pallor, constipation, eructations (indigestion), headache, joint and abdominal pain, weakness in the forearm, stippling of the red blood cells, and weight loss. The predominant symptoms among those defined as poisoned were loss of color (75 percent), stippling (88 percent), headache (52 percent), and indigestion (41 percent).

Second, the incidence of the symptoms provide further evidence that this study likely undercounted the number of cases of water-related lead poisoning. There were 157 respondents who lacked color (pallor), but only 47 of these were classified as lead poisoned. While 82 respondents had a hemoglobin count less than 70 percent, only 25 of these were classified as poisoned. Although 36 of the respondents exhibited vertigo, only 9 of these were classified as lead poisoned. Constipation was also very common; 78 respondents reported irregularity and only 25 of these were classified as lead poisoned.[52]

Wright, Sappington, and Rantoul also presented data about the amount of lead ingested through water. To construct these data, the researchers estimated the daily intake of water for each person in the sample. They based this estimate on direct observation and on information elicited by questioning the respondents. In addition, they took lead readings of each person's water supply. With this information, Wright, Sappington, and Rantoul calculated the amount of lead each person would have consumed through household tap water. Thus 65 respondents (26 percent of the sample) ingested less than 0.1 milligrams of lead per day; 115 respondents (46 percent) ingested between 0.1 and 0.5 milligrams per day; 39 respondents (15 percent) ingested between 0.5 and 1 milligrams per day; 22 respondents (9 percent) ingested between 1 and 2 milligrams per day; and 12 respondents (5 percent) ingested more than 2 milligrams of lead every day. To put this in perspective, the daily dose of Dr. ____'s Famous Female Pills contained 0.2592 milligrams of lead (see chapter 3), suggesting that at least 29 percent of the sample would have been consuming lead levels in excess of those contained in the daily dose of a black market abortifacient. Similarly, a person who drank one liter of tap water per day with the current maximum allowable amount of lead would have been ingesting 0.015 milligrams of lead per day. At least 74 percent of the 239 respondents were ingesting more than 6 times this amount every day.[53]

Breaking the Silence

There were episodes when sudden environmental shocks broke the silence associated with water-related lead poisoning. One of these episodes occurred in Bacup, England, in the summer of 1887, when the town experienced a severe drought. The town's water reservoir went dry for several weeks, and without water regularly running through the distribution system, the interiors of the lead service pipes were exposed to air. This caused the encrustation of organic and inorganic compounds that had built up over the years to break away from the pipes. Prior to falling off, the encrustation had lined the interior of the pipes, forming a protective barrier between the lead and the water. Once that barrier was gone and water was returned to the pipes, water began to dissolve the exposed lead at an unprecedented rate. Subsequent investigations revealed lead levels in household taps ranging from 1/10 of a grain to 2.2 grains of lead per gallon of water, 114–2,511 times the modern EPA standard. These findings were based on a sample of fifty-five households.[54]

Not surprisingly, lead levels like these made many people sick and resulted in multiple adult deaths. Of the 404 cases of water-related lead poisoning documented by Bacup's health officer, 21 suffered from wrist-drop; 197 from headaches; a "few" from "spastic paralysis"; 206 from abdominal colic; 148 from anemia; "several" from insomnia; and 339 exhibited the blue gum line. In 71 cases there was no "patellar tendon reflex"—that is, no knee-jerk in response to stimulation. Vision problems were "not uncommon," including three cases of temporary blindness that lasted from a few minutes to a few days. Vertigo was also "not uncommon" while a "sense of heat and burning in the soles of the feet" was rare but not unheard of. In "nearly every case the perspiration" was "lessened." In 148 cases, "obstinate constipation was present" and the "faeces were hard and dry, and showed deficiency of bile." "A forerunner of saturnine nephritis," albuminuria occurred in "many" individuals. "Shooting pains" in the "hands, arms, back," and legs were "very common," and "usually mistaken for muscular rheumatism."[55]

The most serious cases involved neurological pathologies. Mr. A, aged thirty-five, became homicidal. He "threatened to kill his wife, became very suspicious, and for a time she had to leave him. . . . His expression and mental obliquity were evident to all. . . . He had lead paralysis [and even after treatment did not recover] the use of his hands." Mrs. B, aged fifty-three, "had hallucinations, and did not know her husband. . . . She suffered from epileptiform convulsions and paresis of both hands." Similarly, Mrs. M, aged thirty-eight, was so delusional she "did not know her own family." Mr. J, aged twenty-eight, was paralyzed in both legs and prone to "suicidal mania." J. B., aged thirty-eight, had "epileptic fits [followed] by religious melancholia." Miss J. A., aged twenty-three, while having difficulty swallowing, had "difficulty in speaking [and was] unable to read and sing." Mr. J, aged forty-eight, had "symptoms which threatened to develop into general paralysis of the insane" and was eventually "removed to a lunatic asylum." Mr. J's wife suffered from "glaucoma due to lead." Among those with severe neurological problems, "epileptiform convulsions" were common: there were 24 cases of lead-induced epilepsy and convulsions. Eight of these individuals eventually died.[56]

The drought in Bacup made it easier to diagnose water-related lead poisoning. Because the drought caused lead levels in the water to rise, the severity and strangeness of the associated symptoms could not be ignored or misunderstood. A doctor could easily attribute the odd case of rheumatism, constipation, or abdominal distress to causes other than lead, but it was much harder to mistakenly attribute paralysis, insanity, or loss of sight to other causes. Diagnosing severe cases of lead poisoning was simply too straightforward an exercise. Wrist-drop, delusions, a blue gum line, strange behavior, and bizarre neurological disorders were strong indicators of lead poisoning, especially when they were observed with abdominal distress, constipation, or pallor. Environmental shocks like the Bacup draught generated a sharp increase in the number of cases of water-related lead poisoning—when this occurred, local doctors were confronted with a large number of patients in a very short period time, all with puzzling symptoms. Individually, each of these cases could have been shunted aside and ascribed to nothing more than the odd case of rheumatism, gout, kidney trouble, or some obscure ailment. But when all the various symptoms appeared across many patients simultaneously, doctors were able to identify a constellation of illnesses that all pointed to a single cause.

But even in epidemics of water-related lead poisoning, many cases still went undetected or mistakenly diagnosed. Dr. John Brown, the medical officer for the town of Bacup, believed that the town's epidemic of "plumbism had so simulated other diseases that it had not been recognised even by careful and pains-taking physicians." According to Dr. Brown, there was "no form of disease more insidious, ubiquitous and manifold in its manifestations, and which so closely simulates other diseases as plumbism." In a short treatise on Bacup's epidemic of plumbism, Brown claimed that "scores of cases could be cited in which" water-related lead poisoning had been mistaken for "rheumatism, gout, indigestion, cephalalgia, epilepsy, meningitis, cerebro-spinal-meningitis,

[and] general paralysis of the insane." Brown even believed that "many have died of [these] so called [diseases], which, if traced to their true cause, were really due to lead polluted water."[57]

There is a certain irony to Brown's discussion of mistaken diagnoses because Brown himself appears to have made a critical mistake, or at least a critical omission. In his treatise on the Bacup epidemic, Brown stated that he treated at least 500 people for water-related lead poisoning and he thoroughly documented 404 of these cases. Of the documented cases, however, only 5 were children under the age of five, and only 1 of these childhood cases was discussed in the treatise. The case involved a four-year-old girl who was paralyzed in both legs and suffered from tremors and headaches. Her ailments "simulated anterior-polio-myelitis." Because he was able to document so few cases of childhood lead poisoning, Brown concluded that children were "less predisposed to plumbism than [those] from 15 to 50 years of age."[58] This is perplexing. Modern research makes it clear that lead exposure has more severe consequences for the young than the old.[59] It is also odd that water that contained enough lead to drive full-grown adults insane or into fatal epileptic seizures would have induced so few cases of childhood lead poisoning. Were children in Bacup drinking so little water that they were unaffected by lead levels some 2,500 times greater than the modern EPA standard? Perhaps breast-feeding was especially common in Bacup. But nursing women transfer at least some of the lead they ingest through breast-milk, and recent studies indicate that this transmission can impair infant development.[60] Perhaps Bacup's young drank mainly cow's milk. But cow's milk was routinely diluted with water in the late nineteenth and early twentieth centuries.[61]

Brown acknowledged that lead caused amenorrhea and disrupted the menstrual cycle, but he did not document or mention any such cases in the Bacup epidemic. Similarly, Brown was able to document only 7 cases of lead-related abortions.[62] The absence of more cases of abortion, still-birth, and menstrual abnormalities is surprising. Brown's own estimates of the amount of lead in Bacup's household tap water suggest that the women there would have ingested, on a daily basis, an amount of lead at least 20 times greater than the amount contained in the recommended daily dose of Dr. ____'s abortion pills. (This calculation assumes women were drinking one liter of water a day.) Were pregnant women in Bacup drinking so little water that their developing fetuses were not harmed by the lead it contained? Why would they have consumed so much less water than women who were not pregnant? It is clear from Brown's case studies that Bacup women who were not pregnant were consuming sufficient amounts of water to induce insanity, paralysis, and death. It is odd, that an observer as sensitive to the dangers of lead as Dr. John Brown could not identify the effects of water-related lead poisoning on fertility and the health of the very young.

Alone in a Crowd

An important subtext to Norman Porritt's article was the social isolation he experienced. Because his symptoms were so subtle, Porritt's family and friends did not recognize that there was something physically wrong with him and attributed his gloominess and lethargy to a character flaw. This, in turn, only exacerbated his melancholy and reinforced the opinion that he was a "gloomy person who will not take the trouble to be chatty and lively." Water-related lead poisoning, in other words, not only undermined the doctor's physical health, it also affected him psychologically.

In the case of Dr. Porritt, no one else in his home exhibited any symptoms although they had similar levels of lead exposure. The medical literature of the late nineteenth and early twentieth centuries was full of examples similar to Dr. Porritt. Researchers would describe a particular population where all members of that population were exposed to identical lead levels—these populations usually involved workers in a lead refinery or mine, or animals in a laboratory—yet only a fraction of the population would manifest outward symptoms of lead poisoning. More recent research confirms these crude empirical observations and demonstrates that the ability to tolerate and evacuate lead from the system varies from person to person, and depends upon factors such as genetics, nutritional status, age, stature, personal habits, and overall health.[63]

The irony of all this is that Dr. Porritt's estrangement and social isolation, which were themselves the products of low-grade poisoning, probably exacerbated his underlying pathology. This can be seen in a recent study of lead poisoning among young rats in deprived and enriched environments. The rats in deprived environments were placed in small cages and were isolated from other rats. The rats in enriched environments were placed in much larger cages along with a community of other rats. The enriched environments included stimuli such as, "ladders, boxes, tunnels, wheels, brushes, baby music toys, and platforms." In both the enriched and deprived environments, rats were divided into two groups, with one group drinking distilled water and the other group drinking water containing lead. The lead-treated rats in deprived environments exhibited spatial learning deficits and other markers of neurotoxic exposure, while the lead-treated rats in enriched environments showed smaller deficits in spatial learning. These patterns suggest that deprived environments might exacerbate, while enriched environments might minimize, the neurotoxic effects of lead. When Dr. Porritt gradually withdrew from family and friends and began avoiding enjoyable pastimes such as reading, he might have unknowingly placed himself in a deprived environment, heightening the adverse effects of his lead exposure.[64]

Summary

Building on Norman Porritt's autobiographical account of water plumbism, this chapter has shown how exposure to water lead typically induces subtle and easily misunderstood symptoms in adults. The ubiquity of infectious diseases compounded the difficulties of diagnosing water plumbism. As a result, water-related lead poisoning among adults was underdiagnosed on a wide scale during the nineteenth and early twentieth centuries. A survey of otherwise healthy individuals throughout the state of Massachusetts in 1923 illustrates the pervasiveness of unknown and undiagnosed water-related lead poisoning; this survey suggests that 10–12 percent of the adult population in Massachusetts unknowingly suffered from water plumbism. Only in severe outbreaks of water plumbism, like those observed in Bacup, England, were physicians able to diagnose the pathology on a wide scale. But in places like Bacup, where water lead induced paralysis, insanity, and even death among adults, observers were unable to see the effects water lead had on the unborn and very young.

6

A False Sense of Simplicity

During the 1840s and 1850s, Horatio Adams worked as a physician in Waltham, Massachusetts. In 1852, Adams published a lengthy paper on the dangers of lead water pipes in the journal *Transactions of the American Medical Association*. In the paper, Adams argued that when city officials in the United States and England wanted to install lead water pipes they typically justified their decision by citing two forms of evidence. First, they frequently cited the absence of widespread lead poisoning in cities with similarly situated water supplies as evidence that it would be safe to use lead water pipes in their particular situation. Like some other physicians, Adams believed that this was a dangerous practice because it was so difficult to diagnose cases of water-related lead poisoning if physicians were unattuned to the dangers of lead: "No argument for the safety of lead-transmitted water, founded on the absence of lead malady, ought to be admitted, [because] knowledge of the lead malady [was] so much misunderstood."[1] As shown in chapter 3, Adams views on this score would later be echoed by Norman Porritt and other students of water-related lead poisoning.

Second, as Adams argued, city officials almost universally appealed to the doctrine of protective power to justify the use of lead water pipes. According to this doctrine, over time most lead pipes developed a protective coating on the interior of the pipe, inhibiting the amount of lead subsequently taken up by the water.[2] The use and misuse of the doctrine of protective power had significant public health consequences. Of particular concern was the tendency of many engineers and health officials to suggest that the doctrine implied lead pipes were universally safe, regardless of the chemical characteristics of the water supply in question.

There were three versions of the doctrine of protective power: the Edinburgh doctrine; the Boston doctrine; and the London doctrine. The Edinburgh doctrine was first promulgated by the chemist Christison, and stated that the "right kind" of salts in water produced "compounds of lead known to be insoluble." In this construct, airtight lead pipes exposed to water containing the right kind of salts developed an impermeable coating after a few months' time. Christison identified "the sulphates of soda, magnesia, and lime, as well as the triple sulphate of alumina" and potash, as especially protective. It is notable that Christison was one of the first scientists to suggest that authorities could reduce the lead solvency of water by treating it with chemicals that neutralized the acids it otherwise would have contained.[3] Of the three variants of the doctrine of protective power that will be discussed, the Edinburgh doctrine is the one that has best stood the test of time. Simply put, this doctrine states that hard water is less corrosive than soft water. The reason for this is that hard water contains high levels of calcium and magnesium, which help neutralize the acids otherwise found in water. These elements also help promote the creation of protective coatings on the interiors of pipes.[4]

The Boston doctrine was predicated on the idea that exposing lead pipes to relatively pure water initiated oxidation, whereby an "insoluble coat of suboxide of lead" was created and soon lined the interior of the pipes. This process, it was said, would usually take only a few days—at most, "a few weeks." Adams referred to this as the Boston doctrine because "it was there adopted as the ground of safety of using lead pipe for lake water." Boston drew its water mainly from Lake Cochituate. The proponents of the Boston doctrine maintained that years of experience with lead pipes and above-ground water supplies in Philadelphia, New York City, and London established the accuracy of the doctrine. They argued that there had been very few, if any, documented cases of lead poisoning in these cities—their rarity being due to the formation of a protective coating of suboxide of lead on the interior of the pipes. According to Adams, the central problem with the Boston doctrine was that it was wrong to suggest that the oxidized coating prevented all future corrosion of lead. Even with the oxidized coating, Adams claimed, the water's action on the pipe never ceased and continued to become impregnated with lead, though at smaller levels than would have occurred without the coating.[5]

According to the London doctrine, a small amount of carbonic acid in water prompted the formation of an "insoluble carbonate of lead" on the interior of the pipes. In this way, the London doctrine stated the exact opposite of other variants of the doctrine of protective power. Weak and colorless, carbonic acid is formed by the dissolution of carbon dioxide in water. Adams argued that several prominent chemists in London used this doctrine to justify the proposition that London could have continued to use lead service pipes safely if the city switched to a soft, and more corrosive, water supply. At the time Adams was writing, London drew its water from the river Thames, a hard-water source.[6] Evidence to follow, however, raises serious questions about the London doctrine.

Adams believed that the various doctrines of protective power were correct on some important points. He accepted the idea that certain salts, carbonic acid, and free oxygen influenced the lead solvency of water, but he suggested that there were so many other intervening variables (to be presented here) that one could not predict by measuring the level of these constituents alone, or by laboratory experiments, what would happen in practice. For Adams, the only accurate way to assess a water supply's propensity to take up lead was through experience: put the pipes in the ground; distribute water through the pipes over a long period of time; and measure the lead only after years of use. Because it was so difficult to predict the tendency of any given water supply to dissolve lead, Adams recommended that cities simply abandon the use of lead piping, lest they discover unduly high lead levels after the great expense of installing water pipes had already been incurred.[7]

Adams was not the only observer to suggest caution in trying to predict the solvency of particular water supplies. Even proponents of the doctrine of protective power published articles encouraging doctors to be sensitive to the possibility of water-related lead poisoning among individuals who drew their water from supplies pronounced safe by prominent chemists. For example, in 1860, James R. Nichols published an article in the *Boston Medical and Surgical Journal* arguing that water supplies in Massachusetts were *usually* safe because they contained sufficient amounts of carbonic acid to induce the formation of an insoluble carbonate on the interior of lead pipes. For Nichols, the operative word here was "usually." "But to form an opinion of" the safety of all water supplies, Nichols argued, "we must inquire if the relationship of chemical forces may not be so affected in one locality, as to change the character of the water." Nichols was particularly concerned about how bends and depressions in lead pipe and the presence of organic compounds, such as "fragments of leaves," might influence the ability of water to take up the lead.[8]

Evaluating the Doctrine of Protective Power

Today a large and well-developed literature exists on water chemistry, particularly with regard to its interaction with lead pipes. This literature indicates that Adams and Nichols were right on their main point: There is a high degree of unpredictability in the lead solvency of water supplies. In one computer simulation exercise, it was shown that, in a given water supply, lead concentration levels could vary from as low as 2 micrograms per liter to as high as 80 micrograms.[9] Even in well-controlled experiments on the chemistry of water and lead, it has been shown that small perturbations in a water's chemical and physical characteristics can significantly alter its lead-solvent powers.[10]

What, exactly, drives all of this variability? After water is distributed from its source, the following factors can influence lead solvency: water temperature; age, length, and diameter of the lead service pipe; biological activity within water mains and service pipes (such as decaying vegetation and the development of biofilms on the interior of pipes); decaying lime from cement mortar; and oxidation of large iron street mains.[11] The lead solvency of source water can also vary depending on the season, atmospheric pollution levels, and the presence of biological and chemical agents.[12]

Several examples illustrate the unpredictability of the lead solvency of any given water supply. First, during the 1880s, cities and towns in the north of England experienced a severe outbreak of water-related lead poisoning. The *British Medical Journal* estimated that as many as eight million people might have been affected by the epidemic. The specific counties involved included West Riding of Yorkshire, Lancashire, Cumberland, and Westmorland. Most of the cities and towns in these counties relied on water derived from moorland gathering grounds. Although water from the moorlands had been used for decades without any reported problems, during the 1880s low rainfall caused the water to stagnate and absorb unusually high levels of the peat that covered the moors. Moorland peat contained microorganisms that imparted an acidic quality to the water, making it act "vigorously" upon lead pipes and generating thousands of cases of lead poisoning, many of them fatal.[13]

The solvency of moorland waters might have been influenced by more than just peaty acids. Gilbert Kirker wrote a short article in the *British Medical Journal* arguing that high sulphur levels in the atmosphere might also have played a role. Although Kirker did not refer to it as such, he was talking about acid rain, which is now recognized as a contributor to water-lead levels.[14] Kirker argued that "the products of combustion, which in that part of the country" were "poured into the atmosphere in great quantity and variety" gave the "moorland water supplies" their "abnormal and constant plumbosolvent action." Alfred H. Allen, a chemist, reported experimental results consistent with this argument. Summarizing the results of his experiments in 1882, Allen wrote:

It will be seen that the presence of sulphuric acid, even in very small quantity, notably increases the tendency of the water to act on lead. Repetition of the experiments always furnishes results pointing in the same direction, but the actual figures vary from time to time, being probably influenced by variations in the composition of the water.

Many other writers of the time also believed that coal-burning factories and homes played an important role in the propagation of water-related lead poisoning.[15]

For the scientists who first investigated the epidemic, isolating the mechanisms that made moorland water lead solvent was difficult because it was nearly impossible to replicate in a laboratory setting what happened in nature. As one investigator remarked: "the behavior of these waters on lead is liable to vary exceedingly under varying circumstances, and, unless great exactness be observed, the experimenter may find himself bewildered by the apparently contradictory results he obtains."

Observers warned that the "inferences drawn from laboratory experiments are to be accepted with great caution, unless the experiments are conducted under conditions similar to what obtain in the actual distribution of water." Observations such as these were at the heart of the arguments made by Adams and Nichols regarding the necessity of real-world experience, and not simply laboratory experiments, in ascertaining the safety and lead solvency of any given water supply.[16]

As first suggested by the Edinburgh doctrine, water hardness and alkalinity are negatively correlated with lead solvency; hard water usually dissolves less lead than does soft water.[17] Unfortunately, many historical actors mistook correlation for identification and assumed that hard water never dissolved lead, rather than the more accurate principle that hard water *usually* did not dissolve lead. This mistaken assumption often prompted towns and homeowners to install lead pipes without properly investigating the lead-solvent properties of their water supplies, and to develop a false sense of security regarding their vulnerability to water-related lead poisoning.

To appreciate the significance of this, recall the difficulties physicians had in diagnosing water plumbism when they were unaware of the range of symptoms that could lead to the diagnosis. Imagine then the reluctance if those physicians were not only unaware of the range of symptoms, but hostile to the diagnosis because they knew the patient lived in a hard-water region and were sure that hard water did not dissolve lead. One of the earliest writers to make this argument was the aforementioned James Nichols. Nichols maintained that in "many cities and towns supplied with aqueduct water, physicians not unfrequently meet with certain anomalous affections in patients, which do not readily yield to what seem to be appropriate remedies." Because the physicians are "confident that the general influence of the water [is] harmless...the idea of lead poisoning does not enter the mind, although the diagnostic symptoms point in that direction." Nichols further argued that "the same class of perplexing, persistent symptoms are often met with in individuals and families using well and cistern water, brought to them in contact with lead, and the character of the disease is not suspected until the plumber is required to repair the pipe made leaky by corrosive action."[18] If the homeowners themselves mistakenly believed that their water was not lead solvent, they would have been much less likely to adopt the precautions necessary to prevent lead exposure, such as running water for several minutes before drinking, and installing paper or charcoal filters.

In 1905, John C. Thresh published an article documenting the ability of some hard-water supplies to pick up lead. According to Thresh, it was not "realised generally" that "hard well-waters" in "certain districts" were "capable of taking up poisonous quantities of lead." Thresh went on to document a severe case of lead poisoning caused by hard water. After ordinary use, the water in question contained between 0.3 and 0.65 grains of lead to the gallon (340–740 times the modern EPA standard); after setting in pipes overnight, the water contained between 1.4 and 1.8 grains to the gallon (1,600–2,000 times the modern EPA standard). The mother in this household became ill after two and a half years of consuming this water. She was diagnosed with gout and sent away for rest and recovery. When she came back home, however, her symptoms returned and then worsened, and a year passed before she was diagnosed with lead poisoning. The diagnosis of lead poisoning was made only after the pathognomic blue gum line appeared. By this point, the patient had become "very anemic, suffered from colic and constipation, and finally had intense pain in the occipital region." Other members of the household were also made ill, but not as severely.[19]

It is worth noting that Thresh was not the first scientist to present evidence that hard waters could be lead solvent; Lauder Lindsay had presented such evidence as early as 1859. Nor was Thresh the last. In 1966, a British physician published an article in the *Practitioner* arguing that health officials should not simply assume that because a water supply is hard or alkaline it cannot be lead solvent. As evidence for the proposition, he presented data from a rural English town where more than 10 percent of the local population had been lead poisoned by a hard water supply and lead service pipes.[20]

Similarly, in 1928, two scientists from Illinois took water samples from across the state to assess lead levels. Illinois water was very hard and "so highly mineralized" that analysts had "considerable difficulty in separating the trace of lead present from" the water's many other residual elements but they were able to make some progress. Table 6.1 reports the measured lead levels for several small and medium sized towns in Illinois. Overall, the lead levels in this hard water state were much lower than those observed in Massachusetts, a soft water state, some twenty

Table 6.1
Lead in Illinois water supplies, 1928

Town	Lead, ppm	(Level)/(EPA)	Abort. Equiv.
Champaign-Urbana			
sample 1	0.080	5.3	109.8
sample 2	0.030	2.0	292.7
sample 3	0.050	3.3	175.6
sample 4	0.090	6.0	97.6
sample 5	0.020	1.3	439.1
Pontiac	0.027	1.8	325.3
Decatur	0.110	7.3	79.8
Cedar Point	0.025	1.7	351.3
Stronghurst	0.230	15.3	38.2
Danforth	0.200	13.3	43.9
Chenoa	0.500	33.3	17.6
Norris	0.430	28.7	20.4
Mount Sterling	0.200	13.3	43.9

Source: Rees and Elder (1928), Tables 5 and 7. For calculation of the abortifacient equivalent, see the discussion in chapter 3.

years earlier (see chapter 3). In Champaign-Urbana multiple samples were taken and these ranged from 1.3 to 6 times greater than the modern EPA standard (0.015 ppm). The highest lead levels among the towns sampled were found in Chenoa and Norris, which had water-lead concentrations exceeding the modern EPA standard by factors of 29 and 33, respectively. To get the equivalent amount of lead that was found in the leaden abortion pills discussed in chapter 3, one would have had to consume only 18–20 ounces of tap water daily in Chenoa and Norris. While most Illinois cities had low water-lead concentrations by the standards of the day, the levels in Chenoa and Norris suggest that even in a hard-water region, water sometimes contained enough lead to cause illness.[21]

Picturing the Chemistry of Water and Lead

Like many other New England states, Maine regularly sampled water sources across the state and then published the results in reports of the state health department. In taking these samples, Maine officials mea-

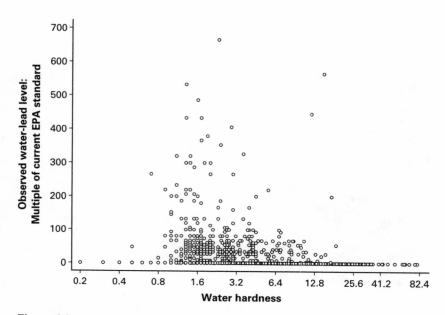

Figure 6.1
Water lead and hardness: Logarithmic scaling. *Source:* Maine State Board of Health (1915).

sured a wide variety of chemical and biological constituents often found in water, including the level of lead, consumed oxygen, water hardness and alkalinity, ammonia, albuminoid, nitrites, and nitrates. Data such as these make it possible to explore the various doctrines of protection and to illustrate their accuracies and inaccuracies.

Figure 6.1 plots the relationship between water lead and water hardness. The hardness measure is an indicator of the amount of calcium and magnesium in a water supply. The figures show that water-lead levels decrease as water hardness rises. The relationship is logarithmic so that, at low levels of hardness, water-lead levels decline rapidly with small increases in hardness. However, there is a threshold effect so that after the hardness measure reaches 20, water-lead levels never rise above 0 and are unaffected by variations in hardness. Furthermore, there is evidence that even at fairly high levels of hardness, water can still dissolve sizeable amounts of lead. For example, one water sample with a hardness measure of 17 dissolved a sufficient amount of lead to place it in

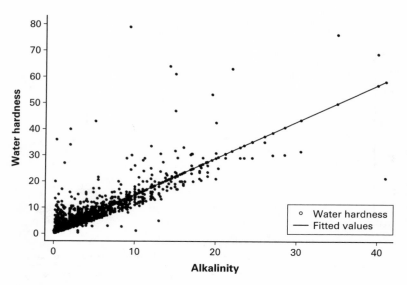

Figure 6.2
Water hardness and alkalinity. *Source:* Maine State Board of Health (1915).

excess of the modern EPA standard by a factor of nearly 600. Any water sample with a hardness measure greater than 14 would have been in the top 5 percent of this sample distribution.

Two inferences might be drawn from figure 6.1. First, as a characterization of a general correlation or trend, the Edinburgh doctrine was correct: Harder water supplies tended to dissolve less lead than soft water supplies. Second, this was only a correlation, and even water that was very hard in relative terms sometimes had the capacity to dissolve lead. This verifies the argument of Thresh and suggests that as the Edinburgh doctrine was elevated to the status of dogma—"hard water never dissolves lead"—it induced a false sense of security among homeowners who drew their supplies from hard water sources.

A water's hardness is correlated with its alkalinity. Hard waters tend to be alkaline; soft waters tend to be acidic. This can be seen in figure 6.2, which plots the relationship between hardness and alkalinity.[22] Alkalinity turns out to be a more reliable predictor of lead solvency than does water hardness, or at least it does for this sample. This can be seen in figures 6.3 and 6.4, which plot the relationship between water lead

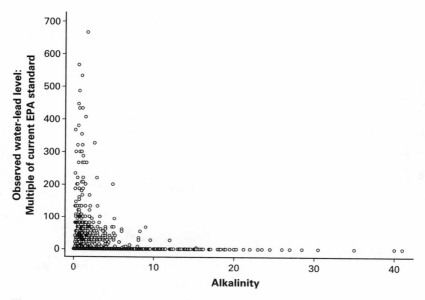

Figure 6.3
Water lead and alkalinity: Ordinary scaling. *Source:* Maine State Board of Health (1915).

and alkalinity. As with hardness, it is an inverse and logarithmic relationship, so at low levels of alkalinity, increased alkalinity is associated with a sharp reduction in the water-lead level, while at high levels of alkalinity, variation in alkalinity has no effect on the lead level. Note that in contrast to figure 6.1, there are no outliers in the data. Overall, these data suggest that if nineteenth-century observers had built a doctrine of protective power around alkalinity, they would have had better success in predicting when and where the use of lead pipes would have been safe.[23]

According to the London doctrine, carbonic acid offered protection against water-related poisoning and it did so by inducing the creation of a coating on the interior of the pipes, as long as the levels of the acid were not too high or too low. Of the three doctrines of protective power identified by Adams, the London doctrine appears the least sound when confronted with systematic data. Using data from Massachusetts in 1899, figures 6.5 and 6.6 plot the relationship between a water supply's

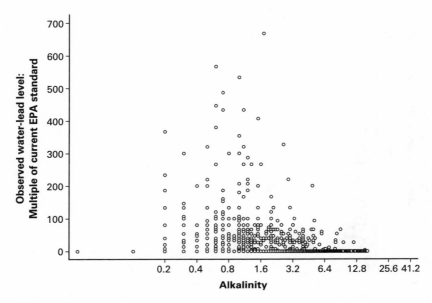

Figure 6.4
Water lead and alkalinity: Logarithmic scaling. *Source:* Maine State Board of
Health (1915).

level of carbonic acid (its CO_2 content) and its water-lead level.[24] Figure
6.5 uses water-lead levels measured after several minutes of ordinary use
(i.e., after the pipes have been flushed), while figure 6.6 uses water-lead
levels measured after the water had stood in the pipes overnight. Neither
figure 6.5 nor 6.6 shows any evidence that, over a relevant range,
increased levels of carbonic acid reduced lead solvency. On the contrary,
the data show that once a threshold has been reached, increased CO_2
levels are associated with increased water-lead levels, and for standing
water, the rate of increase is especially pronounced. Water-lead levels increase more than fourfold once the threshold level of carbonic acid has
been reached.

One objection to the discussion thus far is that the various doctrines of
protective power imply a multivariate model of lead solvency, while the
visual depictions above indicate a bivariate relationship. To address this
objection, a series of regressions are run using the Maine data. The
methods and results are reported in appendix C. The results indicate

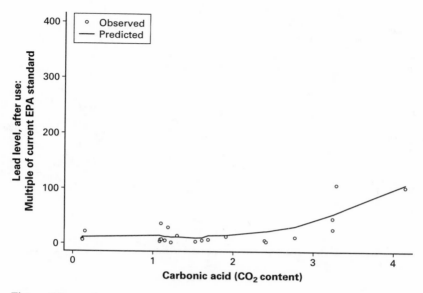

Figure 6.5
Water lead and free-CO_2 levels: After ordinary use. *Source:* Massachusetts State Board of Health (1900).

that hard water and water with high levels of consumed oxygen tend to dissolve less lead than soft water and water with low levels of consumed oxygen. The results also corroborate the idea that there was a great deal of randomness and complexity in water-lead levels; even a fairly extensive regression model explains only 6 to 8 percent of the variation in water lead.

Applying the Doctrine of Protective Power

The legacy of the *doctrine of protective power* was mixed. On the one hand, it helped guide water treatment strategies aimed at reducing the lead solvency in some water supplies. On the other hand, it was applied asymmetrically by cities in their decisions to install lead service pipes. Cities with hard water blindly applied the doctrine to justify their decisions to use lead, ignoring the possibility that hard water sometimes had the capacity to dissolve lead as well. Cities with soft water appear to have

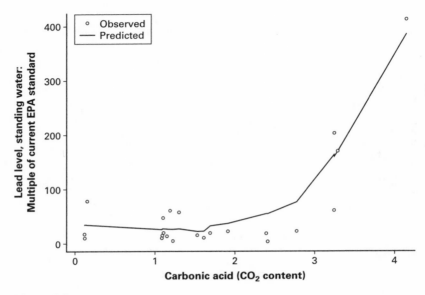

Figure 6.6
Water lead and free-CO_2 levels: Standing water. *Source:* Massachusetts State
Board of Health (1900).

simply ignored the doctrine of protective power, or claimed that the doctrine applied to *all* water supplies.

Consider, for example, William R. Billings, who worked as the superintendent of the waterworks of Taunton, Massachusetts, from 1879
through 1888. In 1898, he published a book on constructing and maintaining urban water systems. According to Billings: "The experience of
every city and town which uses lead for service-pipe is, so far as I can
learn, that a thin brownish insoluble coating soon forms on the interior
walls of the pipe, and then all further action ceases."[25] Put another way,
lead water pipes were safe, no matter what the environmental context or
chemical characteristics of the local water supply. In supporting his claim
that lead pipes were always safe, Billings also incorrectly claimed that
Boston, Worcester, New Bedford, and Fall River used lead pipes yet had
no history of water-related lead poisoning.[26]

Efforts to treat water supplies to minimize their tendency to dissolve
lead were usually a response to experience, though the doctrine of protective power often helped to frame those responses. For example, an

epidemic of water plumbism in Yorkshire and other areas in England during the 1880s drove the affected municipalities to search for methods of preventing water-related lead poisoning. A few towns, such as Oldham, replaced their lead service lines with pipes made of other materials such as iron, or lead lined with tin. These efforts were unsatisfactory, however. Replacement was expensive, and the new pipes, whether iron or tin-lined lead, were acted on by the water as vigorously as were the lead pipes. While tin- or iron-contaminated water was probably not as objectionable as lead-contaminated water, consumers were dissatisfied with the taste and discoloration. In light of these results, most towns chose to treat their water chemically to reduce its corrosiveness. Although treatment processes varied by time and place, they usually involved the addition of calcium, lime, chalk, and other elements to neutralize acidity, and/or hasten the formation of a protective coating on the interior of the pipe.[27]

The idea of water treatment grew out of the Edinburgh doctrine and Christison's discovery that waters containing certain salts and minerals were less corrosive than waters without such constituents. Christison suggested that corrosive water could be made safe by adding these various salts and minerals. The central questions surrounding this idea were: What exactly should be added, and how much? The answers appear to have been location-specific, and depended upon the chemical characteristics of the water supply in question. Only through experimentation and improvements in the understanding of the underlying chemistry of water and lead were water providers able to find reasonably effective and inexpensive ways of treating water to minimize lead solvency. In other words, the optimal water treatment process was not general, but situation-specific, and often required trial and error to arrive at the best-practice technology.[28]

As an illustration of this argument, consider how towns in the north of England responded to the epidemics of water plumbism. In Sheffield, the epidemic prompted local officials to treat their water supplies with lime. Unfortunately, the "addition of lime was generally ineffective" because the limestone was eventually covered with a film that inhibited the interaction of the water and lime. Officials then turned to soda ash, but this "did not give a protective coating which would withstand the attack of the water in the event of accidental inadequate" treatment levels. Finally, the city began adding a small amount of chalk to the water. Although this process was sometimes associated with the development of a bacterial slime, it was inexpensive and generally effective at reducing lead levels.[29]

The town of Guisborough began treating its water supply with lime during the early 1900s. This treatment, however, actually caused an increase in the number of cases of water-related lead poisoning, and officials had to search for alternative modes of treatment.[30] Similarly, in Huddersfield, officials experimented with the addition of silica to hasten the formation of a protective coating. When this did not work, they began using chalk, which reportedly was inexpensive and effective.[31] In Wakefield, officials began treating the town's water supply with sodium carbonate soon after an outbreak of water plumbism. The process was expensive and ineffective. Over the next two decades, local engineers experimented with adding lime and chalk to the water. The central problem was that Wakefield filtered its water through sand filters which altered its chemistry. Eventually it was discovered that the optimal procedure involved adding a small amount of lime to the water *before* filtration, and a small amount of chalk *after* filtration.[32]

Significantly, lead-treatment processes were a fraction of the cost of ordinary water filtration systems, and typically cost only a few thousand dollars for the capital outlay and a few pennies per million gallons of water for operating expenses.[33] In contrast, building water filtration systems to destroy bacterial contaminants such as typhoid could represent 10–20 percent of the total cost of a large urban waterworks.[34]

The Paradox Power

Given the elevated risk of lead exposure in towns with soft and acidic water supplies, one might think that lead water pipes would have been less common in cities and towns with corrosive water supplies than in those with less active waters. The opposite was true, however. The more corrosive a town's water supply, the more likely it was to employ lead service pipes. This can be seen in figure 6.7, which is based on a sample of 130 urban water systems in Massachusetts as of 1905. The x-axis is

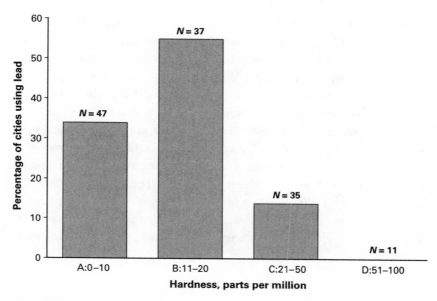

Figure 6.7
Lead use and water hardness: Massachusetts, 1905. *Source:* Whipple (1913), pp. 203–205.

arranged so that the softest and most corrosive water supplies are closest to the origin; and the hardest supplies are the farthest. Notice that of the forty-seven water systems in the state with the softest supplies—that is, those with water hardness between 0 and 10 ppm—around 33 percent used lead service pipes. For the thirty-seven systems with water hardness between 11 and 20 ppm, more than half used lead pipes. However, after water hardness rises above 20 ppm, the use of lead pipes plummets. For the thirty-five water supplies with hardness between 21 and 50 ppm, only 12 percent used lead, and for the eleven supplies with hardness measures above 50 ppm, not a single city used lead.

This pattern is counterintuitive in terms of public-health, but it is sensible in a narrow engineering sense. Waters that acted on lead also acted on iron and various lined pipes, and it acted on those pipes faster than it would have acted on lead. As a consequence, cities that used non-lead pipes in areas with soft and corrosive supplies often found that the pipes burst or otherwise failed within ten years. Replacing service pipes was expensive; the cost of replacing a given pipe was between five and ten thousand in 2005 dollars.[35] Consequently, cities with corrosive water supplies had a strong preference for lead pipes because they corroded more slowly and lasted longer. When iron and galvanized pipes corroded, however, consumers drank more iron and zinc, which were not especially poisonous and might even have been healthful. But when lead pipes corroded, they drank more lead. This sort of nearsightedness persisted even in the face of well-documented cases of water-related lead poisoning in neighboring towns. For example, there was an epidemic of water-related plumbism throughout Massachusetts during the 1890s, yet during the early 1900s more than twenty Massachusetts towns abandoned the use of iron and cement pipes and adopted lead service pipes instead in order to save on the costs of replacing pipes of less resistant metals.[36]

Summary

One might think that the chemistry of water and lead would have been uncomplicated as it seemed to involve only a simple compound, H_2O, and a common element, Pb. During the nineteenth century, many scientists were seduced by the apparent simplicity of this chemistry, and acted on the assumption that it was easy to predict the lead solvency of any particular water supply. This was a dubious assumption that led many large cities astray. As shown previously, the chemical characteristics of water supplies were complex and random. While there were certain characteristics, particularly water hardness, that were correlated with lead solvency, these factors were imperfect predictors of a water's potential to dissolve lead from the interior of water pipes. Paradoxically, cities with the most corrosive water supplies used lead pipes more frequently than cities with non-corrosive supplies because lead pipes better withstood corrosion.

7

Responsibility in the Court of the Absurd

During the nineteenth and early twentieth centuries, judges in England and the United States refused to hold water suppliers liable for outbreaks of water plumbism, forcing consumers to assume the financial and medical burdens associated with water-related lead poisoning. This buyer beware system might have functioned effectively if consumers were well-informed and able to operate without the regulatory fetters of municipal government, but consumers were anything but well informed and unfettered. In particular, consumers depended on their plumbers (who installed and maintained lead service pipes), popular news sources, and local officials to provide them with accurate information regarding the safety of lead service pipes. These sources were typically ill-informed themselves. Moreover, even if consumers had fully understood the potential risks of lead pipes, municipal ordinances mandating the use of lead often prevented homeowners from adopting safer piping materials such as iron and cement.

The High Court

John Jessup Milnes lived in Dalton, a small English town just outside of Huddersfield. He was married, had six young children, and worked as a lawyer in Huddersfield. In July 1881, Milnes was stricken with a "violent colic" of unknown origin from which he quickly recovered. A few months later in September, his illness returned with a vengeance. He was first "attacked [with] wrist drop." Then "he lost almost entirely the use of his arms, his brain was affected, and at one time his life seemed in peril." Milnes was eventually diagnosed with lead poisoning. Although he "partially recovered the use of his limbs," he was still unable "to dress himself" after a year of treatment. After eliminating other possible sources of exposure, the doctors treating Milnes had his tap water tested for lead. Three separate chemists tested the water and found lead levels ranging from 0.34 to 0.84 of a grain per gallon, or 388–958 times the modern EPA standard.[1]

There are two puzzling aspects to this story. First, given the high lead levels in Milnes's tap water, it is surprising that no one else in his home was made observably ill, not even the children, who were all younger than ten years of age. According to one observer, Milnes became ill because "he was a person of a very peculiar constitution." This may have been true, nonetheless lead levels 500–1,000 times greater than the modern EPA standard is a lot of lead. Second, when Milnes sued to recover damages from the Town of Huddersfield (which had supplied his water), he lost, despite multiple appeals. According to the English courts, water companies, whether public or private, were not liable for the damages resulting from lead-solvent water. The consumer assumed all liability.[2]

The rulings against Milnes rested on the interpretation of two statutes, the Waterworks Clauses Acts of 1847 and 1863. According to these statutes, the Town of Huddersfield was "bound [to] provide and keep in the pipes . . . a supply of pure and wholesome water." The word of contention here was "pipes." If the word *pipes* included only street mains and not the service pipes that linked homes and street mains, Huddersfield was in compliance with the statutes, because the water was "pure and wholesome" while contained in the mains, before it entered the lead service pipes. It was only when the water passed through the service pipes that it became impure and tainted with lead. If, however, the word pipes included both street mains and service pipes, it would not have mattered where the water became tainted. Milnes contended that pipes included both street mains and service pipes, while Huddersfield argued that pipes included only street mains.[3]

The courts defined pipes to include only street mains, and ruled that Huddersfield was only obligated to supply "water which was pure and wholesome in the mains." Writing for the majority, Lord Blackburn said that it was unfortunate that Milnes had to "suffer a damage so great without redress," but he did not believe that the town's duty extended beyond the street mains. According to Blackburn, if the water consumers in Huddersfield had imbibed water directly from the street mains, rather than waiting for it to travel through the service pipes, the water would have been safe. Moreover, because consumers owned the service pipes, which were the ultimate cause of the problem, it was the consumers themselves who poisoned the water. "I think," Blackburn wrote, "that the fault, if there was any, was in having lead pipes at all." Lord Bramwell concurred, writing that "the pipe [was] bad, [but] the water good." And the pipe was owned by Milnes.[4]

Two justices dissented from the majority opinion. Their dissents reflected, in part, questions about who really controlled and dictated the use of lead for service pipes. The majority opinion suggested that, because the homeowners owned the service pipes, they had ultimate control over them. This was not the case. Huddersfield forbade homeowners from working on the pipes themselves, adopted by-laws which said that the service pipes "shall at all time be under the control and management" of the town, and mandated that all customers of the waterworks use either lead or cast-iron service pipes. The extent to which consumers had the ability to choose between lead and iron was particularly important for the justices in the minority. In a lengthy dissent, for example, the Earl of Selborne expressed doubt that any real option was given to consumers to use iron rather than lead, and suggested that the use of lead was imposed, de facto, on consumers by the Town of Huddersfield. In response, Justice Blackburn said that he saw no reason to believe that if the "discretion had been left to the inhabitant, there would have been any difference in the material" used for service pipes.[5]

The justices recognized that a decision against Milnes would have far-reaching implications and affect the ability of others harmed by lead-contaminated water to bring suit against public water providers. Lord Blackburn wrote that there were many "very populous districts" that were supplied with "soft water," and in those districts, the "pipes" were "often if not always made of lead." That water in those districts might corrode the lead only showed "how very important the question" before him was. The Earl of Selborne agreed with Blackburn on this point, writing that "the questions raised" concerned "all consumers of water supplied in the usual way by public bodies to inhabitants of large towns, where lead pipes [were] used" and the water had "a quality likely to take up lead."[6]

Milton's Folly

In 1900 and 1901, workers associated with the Boston public works department were installing public sewer lines in Milton, Massachusetts. The lines would eventually be connected to Boston's larger sewer system. In the course of installing the sewers, workers had to drain several private water wells, and left a handful of families without water. As compensation for this, Boston's sewer commission connected these households to the public water system without charge. New lead service pipes were used to connect the households to water mains. During the ensuing eighteen months, cases of lead poisoning developed in eight of the families connected with new lead service pipes; among the eight families, there were seventeen or eighteen documented cases of lead poisoning. Some of the cases were quite severe. There was one death; one young mother went insane; and in several other cases, individuals had been exposed to sufficiently high levels of lead to develop a blue gum line. Tests of the drinking water in these eight households revealed lead levels well above the maximum level then recommended by health officials in Massachusetts (0.5 ppm, 33 times the modern EPA standard).[7]

Significantly, public health officials believed that there were many more cases of lead poisoning in Milton, but that these cases went unreported because physicians and victims incorrectly attributed the symptoms to causes other than lead. Nor did health officials believe that cases of lead poisoning were limited to homes connected with the new lead lines. Lead service pipes were used throughout the town, and about 90 percent of all households with public water used lead pipes to connect to street mains. When the State Board of Health measured lead levels in drinking water in areas with old service pipes they found elevated lead levels there as well, though the levels were lower than in areas with new services. It appears that what enabled Milton doctors to identify lead poisoning among the eight families with new pipes is that they were confronted with a mass of patients in a very short period of time, and these patients' combined symptomatology made lead poisoning the obvious diagnosis. Unfortunately for patients living in areas with older pipes, there was no sudden, epidemic-like outbreak of lead poisoning; there was, instead, a gradual accumulation of various individual-specific symptoms, appearing randomly across patients and time.[8]

One of the families injured in Milton's outbreak of lead poisoning was that of James and Louisa E. Welsh. The Welshes sued the Milton Water Company, a private enterprise, for selling them lead-contaminated water. Mrs. Welsh was directly injured by the water, while Mr. Welsh sued for the "expenses of her illness" and "the loss of her society." At trial, the jury ruled that the water company had been negligent because it failed to test its water for lead solvency, and had thereby unknowingly distributed water that inevitably became contaminated with lead. The Welshes were awarded $4,500 for their pain and suffering. Their victory was short-lived, however.[9]

Sustaining a defense motion, the trial judge set aside the jury's verdict. According to the judge, the Milton Water Company "relied upon the State board of health to notify [it] if there was anything unsafe in the water supplied" to its customers. The water company, however, did not receive any notice of the danger of lead poisoning until after the Welshes had been made ill. Given that the water company did not know that its water would take up lead, and that it had no responsibility to acquire such knowledge, the jury erred when it held the company negligent. This ruling implied that it had been the responsibility of the Board of Health to monitor the safety of the town's water supply, not the responsibility of the Milton Water Company.[10] The Welshes appealed the decision, but to no avail. The appeals court sustained the decision. Although a new trial was ordered, there is no record of the outcome, or whether such a trial ever took place.[11]

The Misbegotten Economy of Lead

The historical accounts recounted here show that before 1930, courts in England and the United States articulated rules that made it difficult for consumers to recover damages for injuries incurred as a result of lead-contaminated water.[12] Such rules created incentives for consumers to protect themselves from harm, because if they did not, the courts would not offer any financial assistance. Due to these incentives, a wide range of household products emerged that were designed to limit the amount of lead in drinking water. Private companies began marketing lead pipes lined with tin and small filters made of paper or charcoal that were attached to kitchen faucets. In spite of what one might expect in an age of patent medicines and diploma-mill medical schools, most of these products were reasonably effective, though they were certainly not perfect.

Tin-lined lead pipes were one of the first products designed to minimize lead in drinking water. As a first-generation product, they left something to be desired. On the one hand, these pipes were nearly as flexible and malleable as pipes of pure lead, but they exposed consumers to far less lead than ordinary lead pipes. On the other hand, tin-lined pipes did not eliminate all water-related lead exposure. As one trade journal explained, these pipes were "lined with tin about a millimeter in thickness, but in the production" of the pipe, the tin dissolved "a considerable quantity of lead." Consequently "the lining [was] not pure tin, but a mixture of tin and lead," and acidic water drawn through these pipes took up a small amount of lead.[13] Moreover, there exists anecdotal evidence to suggest that New York water was sometimes corrosive enough to dissolve the protective lining of tin within a few months and expose the underlying lead, although it is not clear how frequently this occurred.[14]

But the greatest drawbacks to tin-lined pipes were their expense and their tendency to burst, particularly close to joints. One British study found that tin-lined pipe was 4 times the price of regular lead pipe.[15] Due to the expense and uncertainty associated with tin-lined pipes, they never became a popular choice among engineers and plumbers. A survey conducted by the New England Water Works Association in 1917 found that no more than six out of 304 cities surveyed (2 percent) used tin-lined pipes.[16] Such pipes appear to have been used predominantly by wealthy individuals or in large institutional buildings such as schools, jails, and courthouses. In New York City, for example, tin-lined pipes were reportedly used in city parks, schools, hospitals, and various government buildings.[17] In his treatise on eclampsia, the aforementioned

Dr. Porritt wrote that "lead pipes with an inner lining of block tin" were "protective." He reported that he had had such pipes put into his own home in Huddersfield and "found them efficient."[18]

Tin-washed pipes were a second-generation product designed to copy the benefits of tin-lined pipes but eliminate their shortcomings. According to one observer, tin-washed pipes had "proven themselves superior" to lined pipes. Because washing the lead with tin was done quickly, there was no time for the tin to dissolve and absorb any lead. Washing the lead with tin was accomplished by "pouring molten tin into the pipe" as the pipe was simultaneously passed through a die. When the tin passed through the inside of the pipe, it left a thin coat on the inner surface. Although tin-washed pipes appear to have been less prone to rupture than tin-lined pipes, they were not radically cheaper than lined pipes and often the tinned surface was of uneven thickness, affording uncertain protection over the long term.[19]

The most effective domestic product in preventing water-related lead poisoning also appears to have been the cheapest: a small filter made of paper or charcoal that was attached to kitchen faucets. Consider the experience of Sinclair White, the Municipal Health Officer in Sheffield, England, during the 1880s. White ran an experiment in which a solution containing six parts lead per million parts water was passed through seven different paper and charcoal-based filters. "The lead was removed by all the filters except" one.[20] Norman Porritt also advocated the use of domestic filters. "Fortunately there is a ready and inexpensive means of robbing plumbo-solvent water of lead," Porritt wrote. "If the water [is] passed through a filter with a charcoal filtering bed, lead [is] removed from it."[21] Despite these arguments in favor of domestic filters, like tinned pipes, they too were imperfect devices. If, for example, the filters were not regularly cleaned and/or replaced, they lost their effectiveness and perhaps even introduced additional lead into the water.

As explained in chapter 5, Dr. John Brown treated hundreds of cases of water-related poisoning during an epidemic in Bacup, England. In the course of treating his patients, he noticed that those who were the most severely affected were those who were poor and could not afford to purchase charcoal filters for their taps. Through experimentation, Brown was able to design a homemade filter, available for only a "fetching."

His homemade filter system involved running water through a flower pot containing "sand-rock, fine gravel, and sand." Brown claimed that this process removed lead from water "as perfectly and 5 times more rapidly" than household filters sold commercially.[22]

Dr. Brown's invention illustrates an important point. If consumers were so inclined, they could implement their own strategies to minimize the amount of lead in their tap water. Besides making their own filters, consumers could regularly flush their pipes before they drank water or used it for cooking. Although flushing pipes did not eliminate all lead, it did reduce it. One Massachusetts study found that water that had stood in pipes all night contained, on average, three times more lead than water gathered following ordinary use.[23]

But the efficacy of these various practices was really beside the point. Even if these products and strategies were effective and cost only a few pennies, having individual consumers purchase and operate their own lead-prevention systems was neither economical nor effective. Consider a hypothetical city that drew its water from a river or nearby lake, and stored it in a large reservoir near the city. Assume that lead service pipes had been installed in the city many years prior. To prevent water-related lead poisoning, the city could have pursued one of the following three strategies:

• The city could have dug up all the lead pipes and replaced them with pipes composed of a safer material such as iron.

• The city could have hired a chemist to monitor the characteristics of the water in the reservoir, and when the water became lead solvent, the chemist could have added a small amount of lime or some other appropriate chemical to neutralize the water's corrosive properties.

• The city could have adopted a laissez-faire attitude and told all water consumers that they were responsible for preventing lead poisoning, and that if they, or their families, became ill, they would be required to foot the bill.

The first two strategies would have been simple, highly effective and, if done correctly, would have guaranteed the safety of everyone in the city in relation to lead poisoning. The courts, however, encouraged cities to pursue the third strategy.

Under the third strategy, each individual in the city would have had to invest time learning about the problem. They then would have had to identify which strategy they wanted to pursue. If homeowners had doctors with the same dedication as Dr. Brown of Bacup, physicians and the poor would have been running around collecting flower pots and filling them with stones. Imagine a city of ten thousand homeowners—some of them collecting flower pots and stones, others taking time off work to purchase filters, and still others investigating the costs and benefits of investing in a new type of service pipe—all of them duplicating one another's efforts.

Undoubtedly some homeowners would have chosen to monitor the lead levels in their water before investing in any sort of protective device. The simplest way to monitor lead levels would have been to look for symptoms of lead poisoning in oneself and one's family. However, around 1900, there was little appreciation, even among doctors, for just how slow and subtle a poison lead was. If adults waited to observe the overt and undeniable symptoms of lead poisoning in themselves, the damage to their very young children and the unborn would have been significant. Although it does not appear that water companies used anything more sophisticated than observing adult health to monitor lead levels, they, in contrast to consumers, at least had the capacity and resources to regularly monitor their water supplies through chemical analyses. If the courts had held more water companies liable for outbreaks of water-related lead poisoning, perhaps those companies would have been more aggressive in exercising their capacity to test their supplies for lead solvency, and in acting to limit the amount of lead to which consumers were exposed.

As noted in the introduction to this chapter, a system that forced consumers to bear all of the liability associated with water-related lead poisoning might nominally have worked if it were predicated on good information. If consumers were able to discover for themselves with relative ease the possible dangers of lead in their particular locality, they could have chosen to use lead in those environments where water was not unduly corrosive, and avoided lead in those places where water supplies had the capacity to act on lead. Although it is not possible to go back in time and survey water consumers across the world about their knowledge regarding the safety of lead water pipes, it is possible to excavate the informational sources consumers would have consulted to make judgments. Accordingly, the sections that follow examine the advice offered by plumbers, newspapers, and local officials, three sources of information consumers probably relied upon heavily to make their decisions.

Plumb Crazy

Joseph P. Gallagher was a plumber from St. Louis, Missouri. In 1885, at a national plumbing convention, Gallagher delivered a sermon-like defense of the safety of lead water pipes. He began by recounting how, shortly after New York installed its public water system in 1842, "new diseases began to develop" and these new diseases "baffled the most eminent physicians" in the city. "It was taken for granted by many of the physicians that the use of water conducted through lead pipes and used for culinary and drinking purposes was the cause of these mysteries." The plumber had little patience for the views of these "quack doctors," and argued that the true cause of the city's mysterious health problems was "bad drainage."[24]

Based on his own experience, Gallagher claimed that lead pipes were perfectly safe in all environments. His knowledge as a plumber made him certain of his rectitude and perhaps this is what prompted him to use the royal "we." Invoking his own version of the doctrine of protective power, Gallagher argued that an "insoluble film" always developed on the interior of water pipes and prevented the water from taking up excess quantities of lead. Gallagher knew about more than just plumbing, however. He was also a man of letters and religion, a man capable of appealing to God and history as well as science. Quoting the seventh chapter of the Book of Amos, Gallagher sought to link plumbing with God's handiwork.[25] Gallagher also explained how "the terraces of Nebuchadnenezzar's [sic] hanging gardens were covered with sheets of lead."[26]

The point of all this was that, according to Gallagher, lead had been used for thousands of years as a means of transporting water, and yet no evidence had ever been adduced to show that lead pipes were "detrimental to health." "In point of fact," Gallagher said, "there has been no such question asked before our own day and generation." Still using the royal "we," he went on to challenge the doctors who spoke out against lead water pipes but stood idly by as "sixty thousand victims in the United States annually [went] to their graves, poisoned by the use of alcoholic stimulants as a beverage." In light of all the death and destruction wrought by alcohol, how could any sane man focus on something as benign as lead water pipes? "And in the face of these facts," Gallagher railed, "our humanitarians, sanitarians, and philanthropists, come to the front and ask the frivolous question: 'Is Lead as a Conduit for Water Detrimental to Health?'" The answer was an unequivocal, "No!" Lead was "the best, safest, and only material fit for a first class job of plumbing."[27]

Gallagher eventually moved away from history and higher powers and returned to subjects closer at hand. He explained that plumbers generally preferred lead pipes because they were "soft and pliable," could withstand high pressure, lasted "longer than any other material known to the plumbing profession," and were "quickly and cheaply repaired in case of bursting from frost." As evidence of lead's attractiveness as a piping material, Gallagher pointed to Paris, where lead water pipes had been in use for over two hundred years. When these pipes were "taken up" they weighed the same as when they were first put in the ground, suggesting that there had been no external or internal corrosion of the pipes over that time period.[28]

Gallagher's views on the practicality of lead water pipes were typical of engineers and plumbers. For example, an editorial published in *Engineering News* developed many of the same ideas Gallagher had, explaining that lead was "in many respects the most satisfactory" material for water pipes. According to the paper, lead's "pliability" and "comparative freedom from corrosive action" made it "almost ideal from a practical standpoint." Although lead pipes cost more than other pipes, these costs were recouped once one factored in lead's durability and long life span. In the same editorial, *Engineering News* quickly dismissed concerns about lead poisoning, arguing that "lead has always been used for services in most of the large places without any unfavorable effects." While it is true that many large cities did not incur substantial ill effects from the use of lead services, the claim that "most of the large places" used lead "without any unfavorable effects" was questionable even by the standards of 1917.[29]

Nevertheless, both the *Engineering News* and Gallagher were right about one thing. Lead pipes were durable. According to a report compiled by the New England Water Works Association in 1917, lead water pipes typically lasted thirty-five years, and sometimes lasted upwards of 100 years. In contrast, plain iron or steel pipes lasted only sixteen years; galvanized iron pipes lasted twenty years; and cement-lined pipes lasted twenty-eight years.[30]

Gallagher's arguments, stripped of their hyperbole, were widely shared. This should give pause for thought. Consider a homeowner who hired a plumber to install a service pipe to connect his dwelling to a street main or a private well. If that homeowner hired Gallagher, or a plumber sympathetic to his arguments, the chances were good that the service line installed would have been lead. It seems unlikely that Gallagher would have inquired into the corrosiveness of the homeowner's water supply. Consider, too, the possibility of Gallagher testifying before a local government deliberating passage of an ordinance that dictated the use of lead service pipes for the city's water system.

Tin Men

The Colwells, Shaw & Willard Manufacturing Company (CSW) was founded sometime before 1850. Located in New York City, CSW manufactured tin-lined lead pipes for water. As explained previously, tin-lined pipes were said to have lead's durability without the associated health risks. In its marketing campaigns, CSW tried to exploit and encourage anxieties regarding the safety of ordinary lead water pipes. Consider the death of Michael Galler and the subsequent public outcry in New York City. As explained in the prologue, Galler's autopsy suggested that he had been consuming small doses of lead over a long period of time. The medical examiner hypothesized that the ultimate source of this lead might have been New York City's public water system, which employed lead pipes. Soon after these findings were announced, CSW began publishing advertisements like the following in the local press:[31]

AVOID LEAD POISON.—Tin-lined Lead Pipe is the only water-pipe in the market which combines safety to health with strength, durability, facility of bending, and making perfectly tight joints. Water flows through it as pure as if drawn through silver. It is approved by all the leading chemists and physicians in the country; also, the Water Commissioners of New-York, [sic] Brooklyn, and Boston. In addition to the plumbing of houses, it is largely used in conveying water from springs and wells; also, for beer and cider pumps, milk coolers, refrigerators, &c. Circulars and sample of pipe sent by mail, free. Address THE COLWELLS, SHAW & WILLARD M'FG Co., No. 213, Centre st., between Canal and Grand sts., New-York.—*Advertisement.*

Around the same time, CSW also wrote a letter to the editors of the *New York Times* recounting the Galler affair and promoting its block tin pipe: "As the public [seems] greatly excited over the result of the investigation just instituted by Professor Doremus as to the cause of death of Mr. Galler" the question has been raised "as to whether a safer material than lead could not be used for conveying water." After describing the manufacture of its own block tin pipe, CSW explained that other than tin "no other metal, even of the minutest particle [comes] in contact with the water [and its pipe is as] flexible and easy to work with as ordinary lead pipe." According to CSW, their pipe was "strongly recommended by leading chemists and physicians, and also by the Water Commissioners of New York, Brooklyn, Boston, and several other cities."[32]

As noted in the prologue, it was in the 1850s when rumors about lead in New York water first began to circulate. CSW did everything it could to promote and legitimize these rumors. The company even hired a chemist, William H. Ellet, to conduct experiments with ordinary lead pipes and New York City tap water to confirm the danger of using lead pipes with city water. Although he was little more than a hired gun for CSW, Ellet's initial findings were intriguing. His experiments showed, for example, that water standing in city pipes for any length of time contained high lead levels, and that even small disturbances, such as altering the water's chemical characteristics or simply moving or jostling the pipe, could sharply increase the amount of lead in the water. While the press commented positively on these findings in editorials, those comments were brief and Ellet was forced to publish his results as a letter to the editor (as he did at the *New York Tribune*) and as a classified ad (as he did at the *New York Times*).[33]

Ellet then conducted a series of experiments that were more overtly partisan. In the first of these experiments, a piece of CSW's tin-lined pipe was filled with water. According to Ellet, even water that had stood in the pipe for sixty days contained not the "slightest evidence of metal impregnation." If one accepted this finding, the inference seemed "inevitable that tin pipe may be used with absolute safety." Ellet's second experiment compared the capabilities of ordinary lead pipe and CSW's tin-lined pipe to withstand high levels of water pressure. This experiment showed that ordinary lead pipe began to rupture when water pressure was raised to 397 pounds per square inch, while the tin-lined pipe burst only when the pressure was increased to 1,212 pounds per square inch. Again, the results of these experiments were published in local papers as a letter to the editor or as a classified advertisement.[34]

Ellet's experiments drew the wrath of Dr. Meredith Reese, a prominent New York City physician and editor of the *New York Medical Gazette and Journal of Health*. In an editorial laden with sarcasm, Reese attacked the pecuniary motivations of Ellet and the handful of New York doctors who supported his experiments:[35]

The certificates of some half-a-dozen doctors have been marshalled by the chemist against the lead pipe manufacture, and the block tin scrip is said to be rising in the stock market. We earnestly hope that our brethren...who are enlisted in the strife may escape the poison of lead, and be rewarded by their fees in block tin.

Although Reese hoped for a truce with Ellet, he predicted that the truce would come through the disgrace of his fellow scientist:

But a truce to badinage, for never were the brethren caught before in so ludicrous a position, and we opine that they will soon be heartily ashamed of the humbug. That such it is, reason, experience, philosophy and science, attest to all who think and are not paid for their partizanship [sic]...who are entitled to our commiseration, not our censure.[36]

It is difficult to feel sorry for William Ellet. He opened himself up to this sort of attack as soon as he accepted money from a private company to perform what were clearly financially motivated scientific experiments. But Ellet showed as much competence and honesty as anyone could have expected from a hired gun, and his experimental results were not markedly different from what independent and government-sponsored scientists were finding. Ellet was neither the first nor the last chemist to discover undue amounts of lead in New York tap water.[37] Another notable aspect of Ellet's approach was his knowledge of the relevant history and secondary literature. He accurately recounted the events at Claremont, where the French royal family had been poisoned by lead-contaminated water. He also accurately described the experiments of Dr. Christison of Edinburgh, who was one of the first scientists to identify the chemical processes that caused water to leach lead from the interior of pipes.[38]

Although Ellet's pecuniary motivations are not laudable, they are easy enough to understand. The motivations of Dr. Reese, by contrast, are much harder to comprehend. Throughout his editorial denouncing Ellet, he used character assassination to challenge his opponents. At one point, Reese referred to Ellet and his allies (several New York-area physicians thought Ellet was on the right track) as "weak brethren with female nerves." Elsewhere Reese used words like "hypochondriac," "monomania," "mischievous excesses in sensual indulgences," "partisan," and "viscous habits" to describe Ellet and other New York physicians and chemists who claimed to have had acquaintances or patients made sick by lead-contaminated water.[39]

That Reese chose to describe Ellet in these terms is suspect, because Reese's own behavior was anything but unassailable. For example, Reese attacked Ellet because Ellet sold his professional services to promote tin-lined water pipes. The best thing one might say about Reese in this regard is that he was a hypocrite; a more balanced indictment would emphasize the facts that Reese promoted a more dubious set of products than Ellet, and that he abused his position as the editor of an ostensibly professional journal. Consider Dr. Reese's endorsement of the following: Rushton's Cod Liver Oil for the treatment of tuberculosis and other assorted ills; the Hood & Sanderson truss for treating hernias;[40] and a device referred to as an "atmospheric plate" which was used to keep false teeth in place. As published in the *New York Times*, here is how the advertisements for the cod liver oil and the atmospheric plates read:

Rushton's Cod Liver Oil, for Consumption, Scrofula, &c.—Dr. Reese remarks in his *Gazette* of this month, that the name of Rushton has long been identified with Cod Liver Oil, by reason of his laudable efforts to procure and prepare the genuine article for the public, that the son deserves to be sustained in perpetuating the reputation acquired by Rushton's Cod Liver Oil....

A Card.—THE GREATEST IMPROVEMENT YET ACHIEVED IN DENTISTRY.—DR. LEVETT'S PATENT ATMOSPHERIC PLATES FOR APTIFICIAL TEETH.—Dr. Reese, in his *Medical Gazette*, says "many of his patients use them with a comfort and satisfaction hitherto unattainable." The above invention is patented, and the patentee has retained New York City exclusively for his own practice. Applications must, therefore, be made to Dr. Levett, No. 12 Waverely-place, near Broadway.

In both of these advertisements, it is notable that Reese appears to have actively promoted the products in the *New York Medical Gazette*. The advertisement for the truss read:

Rupture.—As the old elliptic Truss, with all its pads and self-adjusting principles, cannot meet the indications of Hernia, the afflicted can learn at 3 Barclay-st., why Doctors Reese, Gilman and Carnochan awarded to HOOD & SANDERSON a gold medal for the best Truss exhibited at the Fair of the American Institute.[41]

Lest one think that the award from the American Institute conferred any independent or third-party legitimation to this particular brand of truss, Reese was the vice president of the institute.[42]

The most curious endorsement offered by Dr. Reese was that for a particular brand of schnapps, Wolie's Schnedam Aromatic Schnapps. "A preparation of juniper berry," the makers of this alcoholic beverage maintained that it had a "delightful aromatic flavor" and "gently stimulating, invigorating, mild, inoxious [sic], and non-intoxicating qualities."[43] Although by the 1850s many physicians were already questioning the medicinal value of alcohol, Dr. Reese belonged to the "old school" and "believe[d] that spirits if kept in proper check" were an appropriate medicine. Moreover, Reese "expressed his decided opinion that 'Wolies Schedam Aromatic Schnapps' [is] the very best and purest article of Holland gin ever produced in the world and a valuable auxiliary in the hands of the medical fraternity."[44]

There is no record of how much the makers of juniper-berry schnapps paid Dr. Reese, but they must have been very grateful. Reese not only claimed that this was "the very best and purest" schnapps ever produced, as editor of the *New York Medical Gazette and Journal of Health* he also dedicated an entire issue of the journal to debate the medicinal benefits of Wolie's Aromatic Schnapps. In using his journal this way, Reese explicitly stated that he wanted the debate to be "open," "impartial," and "free." He did, however, reserve the right to keep the debate limited to "respectable members of the profession."[45]

In the spring of 1855, Dr. Reese published a short article in the *New York Medical Gazette* describing the induction of a new president at the University Medical School of New York City. Reese characterized the outgoing president, Dr. John W. Draper as: "not a medical man"; "had no medical education"; and was "only called M.D., *ex gratia*, by the honorary degree conferred on him by this identical school over which he now presides." These words were not as strong as those Reese leveled at Ellet, but they were serious charges. They were also false. Draper had, in fact, graduated from medical school at the University of Pennsylvania; studied medicine in Europe; and had never received an honorary degree from the University Medical School of New York City. Moreover, when Draper graduated from medical school he "received the extraordinary distinction" of having his thesis published by the University of Pennsylvania. Over the course of his medical career, Dr. Draper had trained almost fourteen hundred students to become doctors. When Draper brought a law suit for libel against Dr. Reese, the latter published a correction in the *New York Medical Gazette*. However, even in this correction Reese never assumed personal responsibility for the mistake; instead he blamed a nameless correspondent for giving him the wrong information.[46]

In July 1853, Reese wrote an editorial (for his journal) in which he denounced the hiring practices of the New York City Hospital. According to Reese, "to secure a berth in that worthy institution, nepotism, family influence, and intrigue are of more value than superior medical knowledge, surgical skill, or large experience."[47] Ironically, five years after he wrote this, Dr. Reese stood in the middle of a much larger and more serious episode of nepotism and favoritism. The episode began when a medical school founded and run by Dr. James McClintock failed. Unemployed and low on money, McClintock launched a new enterprise marketing a wide range of patent medicines that promised to cure everything from whooping cough to malaria. Outraged by his actions, the American Medical Association (AMA) stripped him of his membership. Unfortunately for McClintock, at the same time the AMA ousted him, his patent-medicine business failed.[48]

Dr. McClintock eventually applied for a position as chief surgeon at the Blockley Hospital in Philadelphia. The hospital was impressed with the doctor's previous experience running a medical school, and all that McClintock needed to secure the appointment was the support of a few prominent doctors. McClintock turned to Dr. Reese, who was an old family friend and who attended church with McClintock's brother. Reese happily obliged and wrote his old friend a glowing letter of reference. The reference letter, however, did not mention that McClintock had been expelled from the AMA, and, in fact, intimated that the disgraced physician was a highly respected member of the association. McClintock got the job, but the victory was short-lived. When the medical community in Philadelphia discovered the deception, they were outraged and demanded a public apology from Reese. Reese stonewalled and only issued an apology when the AMA forced him to do so. In the wake of Reese's actions, the AMA altered its by-laws to prevent similar episodes in the future.[49]

Like Joseph Gallagher's paper on the safety of lead water pipes, the Ellet-Reese affair illustrates how difficult it must have been for consumers to acquire safe and reliable advice. Who should consumers have believed, Dr. William H. Ellet, whose financial connection to the CSW company was no secret, or Dr. Meredith Reese, whose own partisanship and prejudice recognized few bounds? That CSW eventually went out of business, and that New York City continued to use lead pipes well into the twentieth century, suggests most historical actors found Dr. Reese the more reliable source of information.

Another Kind of Faith

The *Daily Sun* was the primary newspaper in Lowell, Massachusetts. In 1893, the paper published a short statement proclaiming the universal safety of lead water pipes:

Lead pipe has been traditionally injurious to health from the time of Vitruvius, 2,000 years ago to the present day. In spite of this tradition, millions of people have been drinking water through it from that day to the present time, and it seems to be doubtful...if one well authenticated case of lead poisoning by the use of lead pipes can be found.

In the seven years preceding this statement, the two leading medical journals in Great Britain—the *Lancet* and the *British Medical Journal*—had printed numerous articles documenting thousands of cases of lead poisoning caused by lead water pipes. These cases often resulted in death, paralysis, blindness, insanity, convulsions, miscarriages, and stillbirths. Experts suggested that as many as seven to eight million people may have been affected by lead-contaminated water in England and Wales.[50]

One might defend the *Daily Sun* on the grounds that the events in England and Wales took place in a distant country and were therefore unknown to the editors of the paper. The are two problems with such a defense. First, accounts of water-related lead poisoning in England and Wales had been reprinted in the *Boston Medical and Surgical Journal* and in the journal *Science*. Second, in 1870, the State Board of Health of Massachusetts had launched a survey of water-related poisoning in the state. The board contacted 170 doctors located throughout Massachusetts and asked them the following question: "Have any cases of lead colic or lead paralysis occurred in your town of district, in which you have been able to trace the origin of the disease to water-pipes?" Of the 170 respondents, 41 replied that they had, 109 that they had not, and 20 were not sure. The results were reported in the Board of Health's annual report of 1871.[51]

Sampling a few of the responses received by the State Board of Health in 1870 brings the *Daily Sun*'s editorial into sharp relief. In Ashland, local doctors reported three cases of poisoning caused by lead water pipes. Two of them were a father, aged sixty, and his twenty-four-year-old daughter. The daughter became ill first. She was anemic, had digestive problems including vomiting and nausea, and eventually developed "neuralgia" in the limbs and chest. She visited three physicians. There was some suspicion that she might have been suffering from "a gastric ulcer or carcinoma." Not until her father became ill and developed a blue gum line was the attending physician able to identify a long lead water pipe as the source of the family's strange afflictions. Local doctors reported that upon removal of the lead pipe, "both father and daughter completely recovered."[52]

In Amherst, two cases were reported. One of the cases had all the characteristic symptoms of lead poisoning: colic, constipation, partial paralysis, lead jaundice, and blue line of gums. "Analysis of the drinking water in both cases yielded confirmatory evidence of the presence of lead, and both cases recovered on removal of cause." In Essex, one case was reported. It involved a man, roughly fifty years of age, who had been "subject to attacks of epigastric pain and neuralgia." The cause of his suffering "was not suspected until the extensor muscles of the arm became paralyzed." Only then was it discovered that "he was drinking water conveyed [through] twelve or fifteen rods" of lead pipe. After this practice was discontinued, "he gradually recovered." In Bridgewater, a reported case of lead poisoning involved an eight-year-old boy. After a "gradual decline," the boy developed "epileptiform convulsions, partial loss of speech, [and] power of motion." The cause of the child's suffering "was not suspected for a long time... but when at last discovered and the lead pipe removed from the well, the boy completely recovered."[53]

Epidemic Lead Poisoning in Lowell

Five years after the *Lowell Daily Sun* proclaimed that lead water pipes were safe, Lowell experienced one of worst outbreaks of water-related lead poisoning recorded in modern history. Hundreds of adults and children were made sick. Eventually, the Massachusetts State Board of Health intervened and documented the most serious cases. One adult woman, who had unknowingly been suffering from lead poisoning for two years, was found to have tap water containing 1.1903 grains of lead to the gallon, or 1,357 times the modern EPA standard. She died soon after state authorities documented her case. Another adult female was described by state authorities as a "marked invalid, [whose] fingers contracted on hands and hands on arms." She died from a cerebral hemorrhage shortly after her case was documented. Her tap water contained 0.2891 grains of lead to the gallon, 330 times the modern EPA standard.[54]

Of course, not all cases resulted in death. Much more common were non-lethal symptoms such as colic, constipation, loss of strength, loss of weight, emaciation, headache, and paralysis in the arms and legs. Furthermore, most cases improved once the individuals stopped drinking the city's tap water. After one woman developed paralysis in her hands and wrists, she moved from Lowell to North Billerica and made a "complete recovery." Another woman had "colic and constipation" and eventually became so disabled she could not get out of bed. A nurse who moved into the home to take care of the bed-ridden woman soon developed similar symptoms. Analysis of tap water in the home found that it contained as much as 0.2166 grains of lead to the gallon, 247 times the modern EPA standard. Both women improved after discontinuing use of the city water. In yet another case, a male who lost upper body strength and developed wrist-drop showed "marked improvement" after a "change of drinking water."[55]

These findings leave little doubt of the extent of lead poisoning in Lowell; nor do they leave much doubt about its cause. The Board of Health did not explore the effects of Lowell's lead problem on the unborn and the very young, but given the serious symptoms that had emerged among the adult population it is not difficult to surmise what was happening to the former.

Emaciation, paralysis, even death—one would like to believe that such suffering left the people who ran the Lowell water system remorseful and ready to change their views regarding the advisability of using lead pipes. One's faith would be misplaced. Here is how the superintendent of the Lowell water system, R. J. Thomas, described the state's investigation some twenty years later (emphasis added):[56]

there developed in some sections of the city . . . a number of cases of lead poisoning, *or supposed lead poisoning*. The state authorities investigated it, and they almost issued an ultimatum to the water department that any water that contained over 0.05 of one part in 100,000 parts was dangerous—and there was reputed to be one eighth. I don't know how true that is; *it was reputed to be so*. Although the water department combated that idea, the state authorities thought so. So, about that time, I believe, the State Board of Health doomed this lead pipe.

It is curious that Thomas used the phrase "supposed lead poisoning." Was he suggesting that the many different physicians who diagnosed these cases of lead poisoning were all mistaken? Also puzzling was the claim that the water was "reputed" to contain 1/8 parts lead per million parts water. Was he suggesting that the many different chemists who tested Lowell's water for lead were all mistaken?

Why was R. J. Thomas so reluctant to admit Lowell's use of lead water pipes had been a mistake? Although it is impossible now to go back and probe Thomas's psyche, one possibility seems likely. Thomas was the Superintendent of the Lowell water system during the late 1890s when the epidemic of lead poisoning occurred, and it appears that he had been affiliated with Lowell's water department for many years prior to the outbreak.[57] For Thomas to admit that there was something wrong with Lowell's water would have been a concession that he was at least partially responsible for multiple deaths and serious bodily harm. Admitting to such a thing twenty years after the fact would have been very difficult, particularly when one considers how easy it would have been to fix the problem had Thomas been more forthright while the epidemic was occurring. For example, Thomas could have advised Lowell residents to purchase household filters; to let their water run for several minutes before using it; or he simply could have announced that the water was not safe for drinking or cooking. Alternatively, Thomas could have implemented water treatment to limit the lead solvency of the city's water supply. As explained in chapter 6, water treatment systems were inexpensive and reasonably effective; typically, nothing more was involved than adding a small amount of lime or chalk to the water.

A legal environment that did not force local officials to assume responsibility for their mistakes only reinforced R. J. Thomas's denial.

R. J. Thomas in Context

The behavior of R. J. Thomas was not unusual. Most engineers and water commissioners refused or were slow to acknowledge the problems associated with water-related lead poisoning, particularly when the problems arose in systems they operated or worked for. For example, during the 1880s, Sheffield, England, experienced an epidemic-like outbreak of water-related lead poisoning. In the midst of the epidemic, the engineer of the city-owned water system, Mr. Eaton, referred to "so-called" cases of "lead poisoning" in a report published by the water department. Similarly, at a meeting of the Sheffield Town Council on February 8, 1888, some members of the "Water Committee" said they were "not certain that the lead poisoning was really due to the water."[58]

Alfred H. Allen, who was employed as a government chemist by the Borough of Sheffield and West Riding of Yorkshire, argued that the skepticism of the water commissioners and engineers was entirely un-justified. "This scepticism," Allen wrote, "exist[ed] in face of the united evidence of the medical men that lead poisoning [was] very prevalent" in those areas of Sheffield that were supplied by lead-solvent water. Moreover, local chemists had shown that in the homes of people who were lead poisoned, the water contained anywhere from 0.5 to 1.5 grains of lead per gallon, or 570–1,710 times the modern EPA standard. Allen also wondered how anyone could have denied that the lead poisoning was water related when physicians documented case after case of patients improving once they stopped drinking the city's water.[59]

Historical Antecedents

The decision to use lead service pipes in Lowell was made during the 1840s, decades before R. J. Thomas held any position with the town's waterworks. When Lowell politicians first decided to use lead pipes to distribute water, they were fully cognizant of the propensity of local water supplies to dissolve lead from the interior of service pipes. They had been made aware of this danger by a lengthy report written by prominent physicians. The report presented a series of case studies of Lowell-area residents who had been made sick by lead-contaminated water over the preceding ten years.[60] The city, however, ignored the report and installed lead service pipes anyway.

One example from the long list of cases cited by the report makes clear why local physicians opposed the use of lead water pipes. The case involved a thirty-five-year-old male named E. M. Read. Read was mar-ried, worked as a mason, was "of a fine constitution," and had enjoyed perfect health until he moved to Lowell's Chapel Hill neighborhood in 1836. Soon after moving, his health began to decline and by 1839 the symptoms had become "alarming." The premonitory symptoms were oc-casional attacks of colic, constipation, and a "disagreeable sense of heat in the bowels." He lost "flesh and color, [his] skin assuming a yellowish dingy aspect." Then, pains in his lower limbs developed, along with weakness, stiffness, numbness, "eructations," loss of appetite, and a "va-riety of dyspeptic symptoms."[61]

Other members of the Read family also developed serious illnesses soon after moving to the Chapel Hill neighborhood. Within a year of the move, Mr. Read's two-year-old son "began to be affected with anomalous symptoms in his bowels." The disease was not diagnosed, but it eventually terminated "after two week's illness [in a] fatal disorder of the brain." Although no one was ever able to specify what killed the child, Mr. Read "felt a conviction" that the child died because of the influence of lead in the water of the house. Read maintained that the home's tap water was the child's "constant beverage." As for Mr. Read's wife and only daughter, their health "had become impaired...five or six months previously" to the father's. The wife suffered from "frequent attacks of colic, pain in the limbs, nausea, vomiting, numbness, and stiffness in the lower extremities." The daughter suffered from "spinal irritation, dyspepsia, fetid breath, and convulsions."[62]

The physicians who documented Read's case explained that "similar cases, of a more or less pronounced character," developed in the Chapel Hill neighborhood and other areas of Lowell. All of these cases were eventually "traced to the use of water impregnated with lead." "Not a small number" of these cases "terminat[ed] fatally" either "before the cause was detected, or from an obstinate incredulity on this subject, which prevented its removal at a sufficiently early stage of the affection to admit recovery." Only when this disbelief was "overcome [were] lead pipes generally removed [and] block tin or iron substituted....Forthwith, the intractable character of the disease disappeared; protracted cases began to permanently mend; new ones became less frequent, and finally very rare."[63] All in all, more than six physicians in the Lowell area provided testimony and evidence of cases akin to that of Mr. Read.[64]

When these cases were first observed and treated during the 1830s and 1840s, the individuals affected drew their water from private wells. At this stage, homeowners still had a genuine choice of what sorts of pipe they could use. As the foregoing quotations demonstrate, this freedom allowed consumers to replace their lead pipes with iron or tin pipes once they became aware that the lead was making them sick. However, after the public water system was built, the ability of consumers to opt out of using lead was dictated largely by people like R. J. Thomas.

Self-Denial

The nineteenth-century medical literature is replete with examples of men and women who refused to believe that they were made sick by lead water pipes, even though they and their children had been lead poisoned. In the context of a legal system that placed the burden of prevention on consumers, this sort of denial was problematic. If consumers themselves could not perceive that the use of lead pipes was making them sick, they could not adopt the private strategies necessary for minimizing lead exposure, such as running water several minutes before using it for cooking and drinking, switching to pipes made of safer materials, and installing paper or charcoal filters on household taps. The three examples that follow illustrate the significance of this sort of self denial. All three examples are drawn from Massachusetts during the late 1860s and early 1870s.

In Brimfield, Massachusetts, a local physician reported the case of a man "with the unmistakable signs of lead poisoning." The physician advised the man to stop using "water conveyed through lead pipe." The man, however, "persisted in using it, and finally died unconvinced."[65] In Westminster, Massachusetts, a local physician reported that he had observed "two cases of lead palsy traced to the use of water drawn through lead pipes." One of the cases was "relieved by omitting the water," but the other case was "incredulous as to the cause of his trouble." The man who refused to believe that he had poisoned himself with leaded water was "permanently injured" and suffered from wrist-drop for many years.[66]

In Northampton, Massachusetts, a local physician described an outbreak of water-related lead poisoning in a single family. The family lived in a neighborhood where the residents "very generally" used water "drawn through lead pipes." Because no other family in the neighborhood appeared to have been affected by the lead, the family in question was "not predisposed to accept the theory of poisoning from this source." The father "found himself losing flesh and strength, tormented continually with an unpleasant constriction and pinching in the abdomen and with pains in the extremities." The wife "had similar symptoms." The son-in-law "was still more severely afflicted, being extremely emaciated and feeble." Describing the son-in-law's condition, the attending physician wrote: "His general appearance was like that of one suffering from malignant disease, and without the blue line and family history to aid me in the diagnosis, I should have expected to find a cancerous development somewhere."[67]

The Ideal Consumer

Dr. F. M. Whitsell was a physician in Chicago, Illinois. During the early 1940s, Dr. Whitsell hired a plumber to install a service pipe connecting his home to a street main. The distance from Whitsell's home to the main was seventy-five feet. Whitsell grew concerned when his plumber told him that a local ordinance mandated that a lead pipe be used. Whitsell wondered if there would be "a constant danger of lead poisoning if water used for drinking purposes" was allowed to stand in a pipe seventy-five feet in length. After investigating the situation, the doctor found that the Chicago municipal code "stated that *only* lead pipe can be used in such cases" (emphasis added). Whitsell could not understand why lead pipes "inside the house [were] studiously avoided" while lead pipes outside the house were mandated by law. Whitsell suggested that the municipal code was the relic of a bygone era.[68]

Whitsell eventually wrote to the *Journal of the American Medical Association* and asked the editors their opinion. His letter was published in the journal's Queries and Minor Notes section. The editors solicited the opinions of two anonymous experts in the area, and both of their responses were published alongside Whitsell's letter. According to the first expert, "lead pipe is used in the water supply of many cities and it causes no trouble, because the amount of lead which is absorbed by most waters is negligible." The expert further reasoned that lead piping was "effective in forming an insoluble coating of salts which inhibit[ed] its solution." It was only when "the water supply [was] acid, particularly because of organic acids," that lead was potentially dangerous. Nevertheless, if Dr. Whitsell was concerned about the amount of lead in his water, the expert recommended that he have the water chemically analyzed for its lead content. The expert considered any lead content less than 0.1 ppm (6.7 times the modern EPA standard) to be "perfectly safe."[69]

Whitsell's second expert advisor was also unconcerned about the use of lead water pipes. "The practice of using lead pipe is common perhaps even general in the United States," he wrote. The practice involved "no significant risk to users [because] the quantity of lead...likely to be picked up by water in transit through a short section of lead pipe" was small and tended "to decrease with time because of the gradual deposition of relatively insoluble material on the inside surface of the pipe." The only situations where lead piping might have posed a concern was when the home was new or when the water was highly solvent. Little direct advice was offered to Dr. Whitsell, except for the following remark: "If it can be shown that the lead content of the water in any specific community is unduly high because of the solvent effects of the water on these connecting lengths of lead pipe, there is no doubt that other methods should be employed."[70]

There are two notable aspects of Whitsell's letter and the response it received. First, the experts the AMA consulted both suggested that Whitsell had little to worry about. Lead pipes, they said, were used most everywhere with little evidence that they caused widespread illness. Some fifty years after Joseph Gallagher delivered his bizarre paper on the safety of lead water pipes, the leading journal in American medicine may have abandoned Gallagher's untoward style, but not his essential message: lead pipes were, for the most part, perfectly safe. Second, Whitsell's letter refers to a Chicago ordinance that required all homeowners in the city to use lead service pipes to connect to street mains. The import of this should not be missed. Even Whitsell, who was an especially well informed and savvy consumer, was stuck using a lead pipe because that is what Chicago's municipal code dictated. Such municipal codes appear to have been commonplace in American and British cities, and dated back to the mid-nineteenth century just as Whitsell suggested.[71]

Summary

This chapter has shown that under nineteenth- and early-twentieth-century legal institutions, consumers assumed all financial and personal liability stemming from the use of lead water pipes. This was true even though water suppliers were the lost-cost providers of water-lead abatement. Consumers were often ill-informed about the dangers of lead because their primary sources of information—plumbers, the local press, and public officials—were as poorly informed as they were. Furthermore, even the rare consumer who understood the potential dangers of lead, like Dr. Whitsell, was often prevented from adopting iron and cement pipes because local ordinances mandated lead.

8
The Legend of Loch Katrine

According to nineteenth-century observers, the waters of Loch Katrine stretched "from the Pass of the Trossachs to the moorlands of Stronachlacher and beyond again to Glengyle." Described by some as "the most romantic loch in the world," it was the setting for Sir Walter Scott's lyrical poem, "The Lady of the Lake." "One of nature's most beautiful reservoirs," Loch Katrine was "fed by thousands of trickling rills, meandering burnlets, and scintillant cascades." Although it was surrounded by "thirty-thousand acres of untamed" territory, the loch was only thirty miles northwest of the swarming and industrializing city of Glasgow. "With its congested thoroughfares ... labyrinths of tenement dwellings ... mushroom townlets, [and its many] struggling, despairing, aspiring, and conquering souls," Glasgow stood in stark contrast to the loch. As one partisan observer would claim, the city was a place of "enchainment" while the loch was a place of "enchantment." Yet the two were "inseparably connected, one with the other, [by] the ingenuity of man."[1]

As the foregoing passage suggests, the legend of Loch Katrine had many facets. Some had to do with the wisdom and nobility of the Glasgow politicians who first advocated the creation of the Loch Katrine aqueduct and waterworks; others had to do with the singular purity and softness of the loch's water; others had to do with the economy of the waterworks; and still others had to do with how water from Loch Katrine "magically" reduced death rates in the city. The most significant part of the legend, however, was the belief that water from Loch Katrine would not act on the interior of Glasgow's lead water pipes. Whatever the merits of these claims, the legend of Loch Katrine survived nearly 150 years. It was not until the late twentieth century that the legend gave way to a simpler and less comforting reality—a reality that was not based on the magical qualities of the "most romantic loch in the world." But by then it was too late; the damage had already been done. Admittedly, the damage had been done so quietly and subtly that no one noticed, but it had been done nonetheless.

Before Loch Katrine

During the 1700s, Glasgow had no public water supply to speak of. Instead, local residents relied on public and private wells scattered throughout the city. The wells were located along twenty-four major streets, and in a city park. When people wanted water, they had to walk to the wells and fill buckets with hand pumps. By 1804, a private citizen named William Harley had built a pipeline from a spring he owned in the Scottish highlands. Harley transported pure water into the center of the city, and from there the water was wheeled in cisterns from street to street, where Harley "found eager and ready purchasers of the pure element." Harley charged "one halfpenny per stoup" for the water and purportedly earned £4,000 a year exploiting "the necessity of the people."[2]

In 1801, Glasgow (and its immediate suburbs) had a population of 83,769 people, and the absence of a public water and sewer system was resulting in serious health problems. Butchers dumped carcasses and entrails in the street; city residents did the same with their fetid wastes. As a result, "filth percolated" beneath the cobblestone and earthen streets until few of the city's wells contained anything "other than sewage." "Little wonder, then, that plagues repeatedly visited the city and smote down the poorer inhabitants like flies."[3] Glasgow's high population density compounded the need for public water and sewerage; throughout the early 1800s, there were four to five residents per house.[4] Beyond the public health concerns, there "was also great inconvenience ... much waste of time, [and] no inconsiderable trouble [as] inhabitants from all parts of the city" had to wait their turn at the public wells. An improved public water supply was "therefore urgently required, and loudly called for."[5]

Piped water was first brought to the city with the creation of the Glasgow Water Company in 1806. Two years later, the Cranstonhill Water

Company was formed. Both companies used water from the River Clyde, and located their intake and filtration systems just north of the city. By drawing water from upstream of the city, the companies avoided the pollution generated by city residents. The difficulty with this strategy was that the water had to be pumped from the river to the city and, as a consequence, the pressure was not always sufficient to fight fires. Moreover, because these companies were private enterprises, they installed mains only in those areas that promised the greatest return. The result was neither fair nor economically efficient. In densely populated areas of Glasgow there were two sets of mains when only one set was required, while at the same time, in less densely populated areas of the city there were no mains at all.[6]

In the water industry, which is dominated by high fixed costs and very low marginal costs, economic theory predicts that competition among multiple companies would not be sustainable in the long run. The prediction holds true for Glasgow where, after some twenty years of battle with its rival, the Cranstonhill Water Company stood on the brink of financial ruin. In 1833, the Glasgow Water Company petitioned Parliament for the authority to purchase the Cranstonhill Water Company. The petition was denied because of opposition from the Glasgow Town Council, which believed that higher prices and poorer service would be the inevitable result of a merger. However, by 1838 the Glasgow Water Company had convinced the Town Council that the merger would result in better service and lower prices because a single consolidated enterprise could better exploit economies of scale. The Glasgow Water Company made such arguments credible by agreeing to ceilings on its dividend payments, capital stock, and consumer rates.[7]

Unfortunately, the consolidation of the Glasgow and Cranstonhill companies did nothing to address two enduring problems with the city's water supply. First, as noted above, because the water had to be pumped from the source to the city, water pressure was a constant problem for the city's higher elevations. Second, the water filter, which was a small tunnel filled with sand, had limited capacity. Floods and seasonal changes regularly overwhelmed the filter's capacity and resulted in unfiltered and turbid water entering the mains. As explained by one observer, at an "ordinary state" Clyde water was "pretty free from objection" and "not unpleasant to taste," but during floods the water "was much discoloured by clay" and in certain seasons was "deeply stained with peat." After the Glasgow and Cranstonhill companies merged, the united company made some efforts to increase the capacity of the filtration system, but to no avail.[8]

One of the last areas in Glasgow to receive piped water was Gorbals, the city's southern district. Not until 1846 was the Gorbals Water Company, another private enterprise, formed. The company drew its water from a small stream called Brockburn and two reservoirs located roughly eight miles south of the Gorbals district. Once filtered, the water was piped to the city through a twenty-four-inch main. In contrast to the Cranstonhill and Glasgow companies, the Gorbals company used gravitation to distribute its water and appears to have had fewer problems with insufficient water pressure. The water distributed by the Gorbals company was also "comparatively pure before filtration" and the filters themselves had greater capacity than those used by the Glasgow and Cranstonhill companies. As this brief description suggests, the Gorbals company was a much more effective enterprise than either the Glasgow or Cranstonhill companies had been. Building on its success in Glasgow's southern district, the Gorbals company expanded and started to lay mains north of the River Clyde during the early 1850s. But by this time the company's fate was already sealed; it was clear that politicians and decisionmakers in Glasgow wanted something more than water drawn from mundane streams and reservoirs.[9]

Revolution

At the center of the Industrial Revolution, Glasgow's population grew rapidly between 1800 and 1860. In 1806, when the city's first water company had been created, there were no large factories for spinning cotton and there were no more than 500 textile looms in the city and its immediate vicinity; by 1856, "there were 39 cotton spinning factories and 37 weaving factories [employing] no fewer than 22,455 power looms [and] 27,264 workpeople." In 1806, the city's eighteen pig-iron furnaces produced 22,840 tons of iron annually; by 1851, the number of blast furnaces had grown by a factor of 6, and iron production had

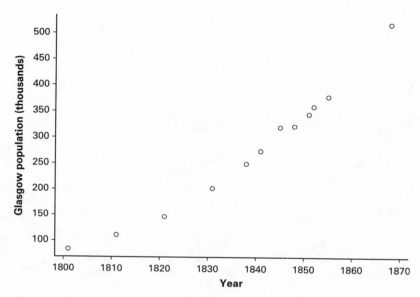

Figure 8.1
The population of Glasgow, 1800–1870. *Sources:* Fraser and Maver (1996), p. 142; Cowan (1840), pp. 260 and 265; Szreter and Mooney (1998), p. 96; Burnet (1869), pp. 15–17; Flinn (1977), p. 302.

increased by a factor of 32. In 1806, there were 24 ships registered in Glasgow ports, carrying 1,960 tons annually; by 1851, there were 508 ships registered in the city, carrying 145,684 tons annually.[10]

The city's rapidly expanding industrial base brought new jobs and new people to the city. As figure 8.1 shows, the city's population doubled every thirty years and increased by roughly fifty thousand people every ten years. In 1801, 5 percent of the Scottish population lived in Glasgow; by 1851, 12 percent of the country's population lived in the city.[11] Glasgow was attractive to migrants not only because businesses there were booming; it was also attractive because conditions elsewhere were so poor. For example, disease, economic stagnation, and famine (1845–1850) fueled emigration from Ireland, and during the early 1800s, the proportion of Glasgow's population that was Irish increased sharply. In 1819, about 10 percent of the city's population had been born in Ireland; by 1831, just under 20 percent of the city's populace had.[12] That so many of Glasgow's immigrants hailed from regions with limited opportunity

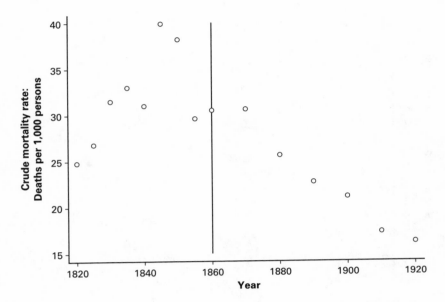

Figure 8.2
Crude mortality rate in Glasgow, 1820–1920. *Sources:* Flinn (1977), pp. 377 and 383; Fraser and Maver (1996), p. 147; Cowan (1840), p. 265.

helps explain why, despite falling real wages in the city, people kept coming. Real wages in Glasgow might have fallen by as much as 15 percent between 1810 and 1840.[13]

Industrialization and uncontrolled population growth overwhelmed the capacity of Glasgow's water system, which, as previously noted, was already heavily taxed. For example, when new factories and dyeworks began sprouting up alongside the River Clyde, they dumped their waste into the river, polluting the water supply. New people and new factories also increased the demand for filtered water, compounding the problems associated with low water pressure. Finally, without any public sewers, Glasgow's new residents did with human waste what had always been done: they dumped it in the street or in shallow pits and cesspools in their backyards.[14]

Faced with a rapidly expanding population and a static public health infrastructure, life in Glasgow became increasingly short and precarious. This can be seen in figures 8.2 and 8.3, which plot the city's crude death

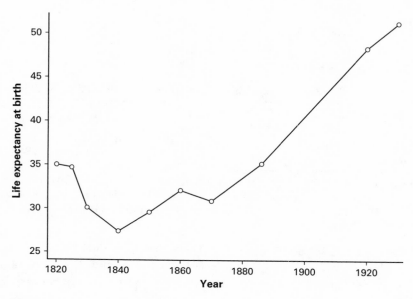

Figure 8.3
Life expectancy in Glasgow, 1820–1880. *Sources:* Szreter and Mooney (1998), p. 96; Leser (1955), p. 71.

rate and life expectancy at birth. The crude death rate rose from 25 deaths per 1,000 persons in 1820 to 40 in 1840, an increase of 60 percent. Life expectancy at birth fell from thirty-five years in 1820 to twenty-seven years in 1840, a decline of 23 percent and a loss of eight years.[15]

The Politics of Cholera

Observers have long attributed Glasgow's many cholera epidemics during the 1830s, 1840s, and 1850s to the inadequacy of the city's water supply.[16] A disease spread mainly by water tainted with human fecal matter, cholera is preventable; it usually cannot flourish in an environment where pure water and public sewer lines are available. Glasgow's worst cholera epidemic occurred in 1832. At least 6,208 residents were stricken, and about half of them (3,166) perished. Another 700 residents, while not killed directly by cholera, were killed by its sequelae or diseases otherwise related to the epidemic.[17] Given that the population of

Table 8.1
Cholera in Glasgow

	Deaths from cholera		Deaths from cholera per 1,000		
Year	Glasgow	Rest of Scotland	Glasgow	Rest of Scotland	Multiple
1832	2,994	7,006	14.8	3.0	5.0
1848	3,892	3,143	12.1	1.1	10.8
1853	3,900	2,100	11.3	0.7	15.6
1866	53	862	0.1	0.3	0.4

Sources: Cowan (1840); "Statistics of the Malignant Cholera in Glasgow, 1848–9," *London Medical Gazette*, 1849, vol. iii, pp. 611–613; Crawford (1854–1855); Flinn (1977), pp. 371–375; Fraser and Maver (1996), pp. 352–356.

Glasgow proper was around 195,000 at this time, these data suggest that 1 out of every 32 of the city's residents had cholera, and that 1 out of every 68 residents died from the disease.[18] For an overview of this and other cholera epidemics in Glasgow, see table 8.1.

In 1848, a second cholera epidemic struck Glasgow. This epidemic was significant in three ways. First, the 1848 outbreak killed 3,892 people, which was a higher absolute figure than the preceding epidemic.[19] Second, the 1848 epidemic was not limited to the city's poor. As described by one early observer, "the striking peculiarity connected with the present distemper, when contrasted to that of 1832, [was that it attacked] a better part of the city and a far larger proportion of the higher and middle classes of the community." "On this occasion" the disease threw "a gloom and mourning over the best habitations of the wealthy" and was "comparatively careless of the more abject inmates of our more crowded hovels."[20] Two members of Glasgow's Town Council even died in the 1848 epidemic. Third, before 1848, people in Glasgow and elsewhere had only suspected that cholera was somehow related to impure water supplies. This changed when the 1848 epidemic struck London and was investigated by Dr. John Snow, who traced the outbreak to a tainted well. Snow's discovery provided additional evidence that Glasgow's water supply was inadequate.[21]

In his landmark book, *Plagues and Peoples*, William McNeill suggested that epidemic diseases like cholera, and the fear that surrounded them, had revolutionary implications for the structure of the state. Few historical episodes better illustrate this idea than the rise of municipal socialism in Glasgow. Municipal socialism refers to the ownership and operation of public utilities, such as gas, electricity, streetcars, and water, by municipal governments rather than private enterprises. In Glasgow— the long-time home of Adam Smith, the arguable birthplace of laissez-faire economics, and a place where the (classically) Liberal party dominated politics for over a century—municipal socialism should have been anathema.[22] But it was not; instead, Glasgow was one of the first cities in the world to municipalize its public utilities.

Prior to the cholera epidemics of 1832 and 1848, there were but a handful of individuals in Glasgow who advocated that the city acquire the private water companies and operate them as a municipal enterprise. These people were an odd mix of religious and civic leaders who perceived themselves as "the natural guardians of the interests of the community"—industrialists who used steam power in their factories, and radical labor leaders who saw municipal socialism as the necessary first step in a broader societal transformation.[23] But with each cholera epidemic, the idea of public ownership gained broader support. One year after the 1832 epidemic, a local politician introduced a bill that would have enabled the city to purchase the Glasgow and Cranstonhill water companies "provided they were willing to sell on reasonable terms." When the companies declined, the matter was dropped. After the 1848 epidemic, a similar measure was introduced, and again the companies declined. This time, however, the matter was not dropped and the city began making plans to build a municipally owned water system to compete with the private companies.[24]

Although the city's efforts to build a municipally owned waterworks did not pass without a fight, the city ultimately won the battle and by 1855 Glasgow had a municipally owned water system. The creation of a municipal waterworks broke down the ideological and economic barriers that would otherwise have hindered the creation of subsequent municipal enterprises in the city. As Hamish Fraser and Irene Maver said in their recent history of Glasgow, municipalizing the water system "helped to legitimise the municipal option, given that market forces had not proved a successful means" of governing the city's water industry.[25]

The municipalization of the city's water supply in 1855 was followed by municipal takeovers in electricity (1890), gas (1869), and streetcars (1872). During the late nineteenth century, Glasgow politicians went so far as to municipalize part of the city's housing stock in an effort to improve the lot of the poor.[26] For the advocates of municipal socialism, Glasgow was a model and an inspiration.[27]

Building the Aqueduct

By 1850, it was clear to everyone in Glasgow that the prevailing water situation was untenable. The River Clyde was heavily polluted, and it appeared impossible to maintain adequate water pressure. A new water source had to be found. Entrepreneurs and private enterprise offered a plethora of options, some good, some bad. In 1836, an engineer from Rothesay proposed bringing water from the North Calder and the Avon rivers to Glasgow. The plan was not adopted. In 1837, a Mr. Sirrat of Paisley proposed a scheme for damming up the Rowbank and Cowden-mill Burns. It too failed to generate support. In 1838, there were proposals to bring water from the Calder, Loch Lomond, the Allander, and the Endrick. "All were advanced as sources and immediately rejected."[28] In 1849 and again in 1852, the Gorbals Water Company proposed impounding water from the River Cart and its tributaries. In one of the more ambitious proposals, the Glasgow Water Company suggested running a pipeline to Loch Lubnaig, roughly thirty miles north of the city.[29]

In November 1851, Dr. W. J. Macquorn Rankine and Mr. John Thomson suggested building an elaborate aqueduct and pipeline to Loch Katrine. Their plan was compelling and based on a systematic examination of all available alternatives. Rankine and Thomson wanted to identify a water source that best combined "the advantages of abundance and purity [with] facility and security in construction of the works, and economy in their execution, maintenance, and management." They directed their attention to a "range of hills to the north" of the River Clyde. Drawing water from hills would create high water pressure through gravitation, and would thus eliminate the need to pump water from a lower elevation. In addition, water sources in these hills were "exceedingly soft and free from mineral impregnation, [and] the quantity of organic matter in the water [was] imperceptible." (As noted previously, the Clyde was regularly polluted with peat, mud, and decaying vegetation.) Finally, "the entire absence of mines" in the area removed "all apprehension of danger to the works from subsidence."[30]

Having chosen the general area from which to draw the city's water supply, Rankine and Thomson then narrowed the prospective sources down to two large, natural reservoirs, Loch Lubnaig and Loch Katrine. Both of these sources were at an elevation some three hundred feet higher than Glasgow, and both were "large lakes of extreme purity." Of the two sources, Rankine and Thomson reasoned that Loch Katrine was preferable because Loch Katrine was at a higher elevation, which meant that for a pipeline of any given size, "a greater quantity of water" would be "discharged." Geography made it easier and cheaper to build a pipeline to Loch Katrine than to Loch Lubnaig—the latter would require extensive tunneling, including a tunnel three miles long near Milngavie that could not have been completed in less than five years. Finally, the "natural outlet of Loch Katrine" made it easy and economical to maintain appropriate water levels; to accomplish the same thing at Loch Lubnaig would have been "very difficult and expensive."[31]

Rankine and Thomson were particularly impressed with the biological and chemical characteristics of Loch Katrine water. "As to purity and softness" the water was "unparalleled." It came from "streams rising in a district of mica slate, almost entirely uncultivated and uninhabited." The water was "at all times so clear that filtration might be dispensed with, even after the greatest floods." The water was also extremely soft, containing only two grains of solid matter to the gallon, and scoring less than one degree on the hardness scale. In this way, the water from Loch Katrine surpassed the "celebrated water of Aberdeen."[32]

Rankine and Thomson created a joint-stock company in an effort to raise the capital necessary to build the Loch Katrine aqueduct and the associated distribution system, but in light of the competition from two incumbent firms, few people were willing to invest in such a large and risky venture.[33] Unable to raise the money in private capital markets, the two visionaries wrote a lengthy letter to the Glasgow Town Council recommending that the council take stock in the company and operate it as a municipal enterprise. Although Rankine and Thomson were disappointed that the venture would not be their own, they maintained that the city would probably be better served by a municipal enterprise anyway.[34] Greeting this proposal with great enthusiasm, a member of the town council made a motion to purchase shares in the company valued at £10,000, provided that, among other things, three members of the town council "shall be *ex officio* directors of the Loch Katrine Company." This was too fast for other members of the council, and several councillors suggested that a committee should be appointed to fully investigate the advisability of the Loch Katrine scheme.[35]

The subsequent investigation confirmed the findings of Rankine and Thomson regarding the superiority of Loch Katrine over Loch Lubnaig. This finding spelled certain death for the future of Glasgow Water Company and its plans for future extensions. The investigation, however, gave lukewarm approval to the plan of the Gorbals Water Company to exploit the River Cart and its tributaries. According to the investigating committee, "the capabilities and proposed extensions of the Gorbals works [were] in every way worthy of attentive consideration." The company's filtration system was described as "ingenious" and "sufficient." Water in the Gorbals system, while not as clear and pure as water from Loch Katrine, was the best available from sources south of the city. The strongest argument against the Gorbals plan was its limited capacity. When finished, the Loch Katrine plan promised to deliver the city at least 50 million gallons of water a day; the Gorbals plan could promise no more than 20–30 million gallons daily. Local politicians reasoned that if the city continued to grow as rapidly as it had in the past, it would soon outgrow the capacity of the Gorbals plan. They therefore "unhesitatingly preferred" the Loch Katrine plan. As figure 8.1 shows, the city's subsequent growth trajectory vindicated this reasoning.[36]

Once the investigation was completed, Glasgow politicians secured the legal changes needed to acquire the works of the Glasgow Water Company and the Gorbals Water Company. After acquiring the works of the Glasgow Water Company, the city continued to use the company's distribution system but cut off the use of the Clyde water when water from Loch Katrine became available in 1859. Water from the original Gorbals system, however, continued to be used mostly by residents in Glasgow's southern districts. As late as 1868, 14 percent of all water consumed in the city was derived from the original Gorbals sources. The continued use of the Gorbals water is an indication that this water was as safe and free of disease as was water from Loch Katrine, even if there was not enough of it to supply a rapidly growing city.[37]

Work on the aqueduct linking Glasgow to Loch Katrine began in May, 1856, and was completed three-and-a-half years later. The aqueduct was thirty-four miles long. It included nine tunnels, two of which were more than one mile long, and numerous shafts 50–160 yards deep. The tunnels were eight feet in diameter, and dropped about ten inches for every mile. Much of the tunneling required blasting through hard rock such as mica slate and quartz. In an area that was particularly dense, "the progress did not exceed three lineal yards per month, although the work was carried on day and night." The aqueduct also included twenty-five iron and masonry bridges built over rivers and ravines; five of these were between 134 and 332 yards long. More than three thousand laborers worked on the project, five of whom were killed in accidents, which one observer found surprisingly few, given the "dangerous nature of the work." All told, the aqueduct cost £468,000 as of 1859, or about £29,218,600 in 2002 pounds.[38]

Making a Legend

Robert Crawford was a member of the Glasgow Town Council, a long-time advocate of public ownership, and, for at least ten years, served on the committee which oversaw the operation of the city's municipally owned street railways. In 1906, Crawford published an article in the *Annals of the American Academy of Political and Social Science* extolling the virtues of municipal ownership in Glasgow. Crawford argued that Glasgow had "for many years taken a wide and comprehensive view of the work which its governing body should carry out for its inhabitants." According to Crawford, no city in the United Kingdom could claim a "broader and more extensive" set of "municipal operations." Not only did the corporation of Glasgow "undertake all the duties which naturally and necessarily [had] to be discharged by every city government, [it also handled] large enterprises of a commercial character [that were more] commonly entrusted to private enterprise." The citizens of Glasgow were so satisfied with the city's operation of gas, electricity, water, and street railways that there was "probably no citizen of Glasgow" who was "so foolish or so bold" as to propose "that any of these four great natural communal enterprises should be divorced from public control and handed over again to private ownership." "No sum of money," "Crawford continued, however extravagant, and no conditions however apparently advantageous" would have been sufficient "to tempt the people of Glasgow to part" with municipal ownership.[39]

Crawford's discussion of the city's municipal water system is illustrative. He described the city's water supply as "remarkably uniform in quality, temperature, and color," and "absolutely free from pollution." "There was," according to Crawford, probably "not another city in the [United Kingdom or] anywhere on the face of the earth [where a] large population [was] so advantageously situated as the people of Glasgow for water supply in respect of abundance, cheapness, and purity." The value of the city's water system was "of incalculable value in fostering the arts and industries of the city, [and its] effect on the tables of mortality [was] something magical." Given all this, Crawford doubted that the "whole effect of this great and bold municipal venture [could ever be] reckoned up."[40]

Crawford was not a lone voice. On the contrary, the historiography of municipal ownership in Glasgow is dominated by this sort of unqualified enthusiasm. This has resulted in an unbalanced portrayal of the successes and failures of the city's public services by more recent scholars. Consider the portrayal found in the authoritative history of Glasgow by Hamish Fraser and Irene Maver, where the Loch Katrine waterworks are characterized as a "spectacular reality." Although this reality did not develop unopposed, it ultimately triumphed: "The route to Loch Katrine had proved arduous, given the vested interests and ideological scruples which had to be overcome before wholesale municipalisation could be adopted." Nevertheless, once "councillors committed themselves, Glasgow's publicly owned water supply was identified irrevocably with benevolent civic interventionism, becoming an inspired and enduring public relations motif for the city." And if this were not enough, "not only did Loch Katrine embody the nurturing quality of pure Highland water, but Sir Walter Scott's epic poem *The Lady of the Lake* was set amidst its romantic landscape.... The loch was thus imbued with symbolism for Glaswegians, not least because its use as a water supply was perceived as restoring nature's balance in a dangerously disjointed society."[41]

Perhaps an even clearer example of this lack of balance is a short book by the Glasgow politician John S. Clarke. Published in 1928 and titled *An Epic of Municipalisation: The Story of Glasgow's Loch Katrine Water Supply*, Clarke's book was reprinted from articles that had originally appeared in the *Forward*, the self-proclaimed "great socialist weekly." Here is how Clarke summarized the history of the Loch Katrine water supply:

Let us then feel proud of the singular privilege we enjoy, and honour the memory of those who first conceived the scheme, battled so tenaciously for it, and finally triumphed. Beside their achievement, a romance of selfless devotion to duty and great purpose nobly accomplished, the other romance of Loch Katrine withers into insignificance.

After announcing that his own epic story would satisfy "the romantically inclined as well as the more prosaic," Clarke lamented the many Glasgow residents who had forgotten the men who first built the Loch Katrine aqueduct. He took heart, however, in the few who appreciated "the doers among the sons of men who have striven to assist humanity during their temporary sojourn on this globe of struggle"—the few who carried "in their hearts a gratitude for the splendid services" provided by the architects of the Loch Katrine aqueduct.[42]

If Clarke's language sounds excited or strained, perhaps he should be forgiven, for he viewed the political battle to municipalize Glasgow's water system as an epic struggle between the forces of good and evil, between the forces working for justice and those working for greed. That Clarke felt this way is especially clear from the following passage, which denounces the fact that the City of Glasgow had to compensate the Gorbals and Glasgow water companies when it acquired the capital of both companies:

This beautiful system is called Capitalism. A better name for it is Parasitism, and when you grasp exactly what it means to the perfectly functionless drones who draw the interest year in and year out, do you wonder that every effort made to stop it is met by these people with vicious slander and downright lying?[43]

What was the "vicious slander and downright lying" to which Clarke was referring? Partly it appears to have been a reference to the opponents of municipal ownership who characterized municipal socialists as "atheists, free lovers, birth-controllers," or any other name that aroused "prejudice." But more importantly, it was a reference to the systematic and manifold efforts of the private water companies in Glasgow to block municipalization.

According to Clarke, the Glasgow and Gorbals water companies marshaled "every reactionary battalion" for their cause. The "trump card" of these reactionary battalions was the claim that water from Loch Katrine would leach lead from the interior of the city's water pipes and thereby poison local residents. As portrayed by Clarke and subsequent writers, the logic behind this trump card was simple: If the water companies could convince city residents that water from Loch Katrine was dangerous, it would have made their own alternative plans, to pipe in water from Loch Lubnaig or from the River Cart, that much more attractive. Observers like Crawford and Clarke dismissed the claim that water from Loch Katrine could become impregnated with lead as a fiction born of greed and whole cloth. Alluding indirectly to the lead-poisoning concern, Crawford wrote: "Every conceivable kind of calamity was predicted" and "the most absurd criticism was indulged in." Clarke was much less reticent, arguing that one of the chemists employed by the water companies to measure lead levels "had been engaged to do a shady job," although in the end, "he was not crooked enough to do it thoroughly."[44]

The name of the "shady" and "crooked" chemist was Frederick Penny. He worked as a consultant to the Gorbals Water Company, but his primary employer was Andersonian University in Glasgow, where he served as a professor of chemistry. During the years 1853 and 1854, Dr. Penny ran a series of experiments designed to assess the lead solvency of water from Loch Katrine. In one experiment, he removed a lead service pipe that was ten years old from a tenement house in Glasgow. He then filled the pipe with water and measured how much lead was dissolved from the interior of the pipe. "In 24 hours this pipe gave 1/5 grain per gallon, and in four days half a grain per gallon." Experiments on new pipe yielded lead levels between 1/40 of a grain per gallon and 3.5 grains per gallons, or 28.5 to 3,990 times the modern EPA standard. Penny examined the possibilities that the level of light and oxygen, whether the vessels were open or closed, and the quality of the lead pipe might too have influenced the amount of lead taken up by the water. These experiments also yielded evidence of undue lead levels. Although the Loch Katrine water was the "purest and softest water" Penny had ever examined, he concluded that the water acted "powerfully and to a dangerous extent on commercial lead in all its forms, both corroding and dissolving the metal."[45]

Clarke maintained that Penny was a hired gun who cooked his data for the benefit of the Gorbals Water Company. If this was so, it is difficult to make sense of the following passage with which Penny concluded his first report to city officials:

For my own part, I believe it quite possible to bring Loch Katrine water to Glasgow as pure and soft as it is in the Loch; and, on the other hand, I think it would be quite easy, by doctoring it on its journey, as proposed in the Promoters' Evidence, to change its qualities completely, and, by increasing its degree of hardness, to deprive it of its present vexatious power of corroding and dissolving lead.[46]

In short, Penny did not recommend that the city abandon its plan to use Loch Katrine water; he only suggested that the water be treated to minimize its lead solvency, a suggestion that had also been made by the original promoters of the Loch Katrine scheme. That the original promoters of the Loch Katrine aqueduct recognized the possibility that water from the loch might dissolve undue amounts of lead suggests that Penny might well have been on to something. Moreover, as explained in chapter 6, water treatment processes designed to limit lead solvency were fairly cheap; the processes typically involved little more than adding a small amount of lime or chalk to the water.

Of course, those who favored municipal ownership hired their own experts and chemists to challenge the work of Professor Penny. Building a model that was designed to mimic the ultimate Loch Katrine aqueduct, these chemists examined the lead solvency of the water after it passed through the model aqueduct and distribution pipes. This approach was taken because many observers believed that the lead solvency of water was altered by contact with other metals and stones. The upshot of the many experiments was that, after passing through the aqueduct, Loch Katrine water lost its ability to dissolve lead outside of a very small window. Specifically, the water acted vigorously on the lead during the first few minutes of exposure, but the lead was quickly covered by an impermeable layer of oxidized material. There was speculation that Loch Katrine water lost its ability to act on lead because the aqueduct, both the model version and the real-life version, included several tunnels lined with limestone which altered the water's chemical properties.[47]

For many of the chemists, the strongest evidence against Professor Penny was the historical experience of cities other than Glasgow. For example, Professors Thomas Graham and A. W. Hoffmann wrote: "We would press strongly the facts that the water of Loch Katrine is in no respect peculiar or exceptional in its composition and properties, and that the safety for town use of the class of waters to which it belongs has already been decided by the most ample experience." They cited, in particular, the cases of Boston, New York City, and Philadelphia. Each of these cities, they claimed, had water as soft as that from Loch Katrine, yet physicians in these cities testified "that no case of lead-disease from this cause has ever been heard of." Graham and Hoffmann also had occasion to examine the waters of Inverness, Scotland, and Whitehaven, England. Water in these towns had "the same degree of softness as the water of Loch Katrine, and also the same decided action upon lead.... Yet the injurious action of the metal upon the water in use in these places [had] never been observed nor even suspected, [nor had] a trace of metal [ever been] found in the water."[48]

This argument was made repeatedly in the reports of the various experts. Table 8.2 lists all of the larger towns that were cited as having water similar to Loch Katrine's and that used lead pipes and cisterns.[49] According to experts, the water in two-thirds of these places "had a very sensible action" on lead. In most cases, the action was equal to Loch Katrine's, but in a few the action was greater. Although these cities all used lead pipes and lead-solvent water, "no single instance" of a "trace of lead" in the water had been discovered, and local doctors reported that they had "never known a case of injury arising from the supply of water."[50] Table 8.2, however, also reports the history of these towns after 1854. Of the thirteen cities listed, only five had no

Table 8.2
Cities with purportedly lead-free water supplies

	Post-1854 history of water lead	
City	Excessive lead level	Cases of lead poisoning
Scotland		
Aberdeen	No history	No history
Edinburgh	Yes	Yes
Perth	No history	No history
England		
Blackburn	Yes	Yes
Bolton	Yes	No history
Manchester	Yes	Yes
Rochdale	Yes	Yes
Sheffield	Yes	Yes
Wales		
Bangor	No history	No history
Beaumaris	No history	No history
United States		
Boston	Yes	Yes
New York	Yes	Isolated cases
Philadelphia	No history	No history

Sources: Allen (1888); Brown (1889); Greene (1889); Hills (1894); Ingleson (1934), pp. 55–68; "Lead in Lancashire Water," *Lancet*, July 11, 1908, p. 120; Local Government Board (1888–1889, 1893–1894); Massachusetts State Board of Health (1899, 1900); Quam and Klein (1936); Thomson et al. (1989); "The Water-Supply and Lead Pipes," *Lancet*, April 24, 1909, p. 1212. For New York City, see the prologue. For Aberdeen, see Smith (1852).

subsequent history of water-related lead exposure, while eight of the cities—Blackburn, Bolton, Boston, Edinburgh, Manchester, New York, Rochdale, and Sheffield—had documented histories of undue lead exposure in the years following this investigation. It should also be noted that while Aberdeen is not included in this list, the water there contained lead levels as high as 1/20 of a grain per gallon, 57 times the modern EPA standard.[51]

The promoters of the Loch Katrine plan also used a misleading standard with regard to what constituted a safe level of lead in the water. In particular, the promoters stated that one grain of lead per gallon of water was "the smallest proportion...known to produce injury to health." This was incorrect. Only a few years earlier, Dr. John Smith had published a paper showing that as little as 1/20–1/10 of a grain per gallon had a "manifestly deleterious action." Similarly, there was a widely cited case of a small river town where the inhabitants had regularly consumed 1/9 of a grain per gallon and had suffered "deranged" health. Putting all of this in a modern context, the promoters of the Loch Katrine scheme were maintaining that a lead level that would have exceeded the modern EPA guideline by a factor of 1,140 was perfectly safe. Dr. Penny was skeptical of this standard and wrote that "few who have considered the importance of this question would willingly partake, year after year, of water containing even 1-10th of a grain of lead per gallon." Penny further argued that to many observers, "all lead contamination is objectionable [and that] no degree of it is safe."[52]

In the court of public opinion, Penny's arguments were roundly defeated. His simple experiments did not compare with the elaborate model of the Loch Katrine aqueduct constructed by the rival scientists. His scientific modesty suggested to at least some observers that he was incompetent. Penny, for example, said that he did not view his experiments as the last word on the subject, and asked that "more experiments and enquiries be made" before a "final answer" was given. Penny also repeatedly stated that the water of Loch Katrine was very pure and did not require any filtration before distribution. According to the municipal-socialist John Clarke, the event that "turned the tables" on Penny and the other "water profiteers" was the publication of a front-page article in the *Glasgow Mail* sometime in 1854. This newspaper article emphasized that lead pipes had been used for over half a century in Glasgow without any ill effects. If there were no problems before, the paper reasoned, why should there be any after the introduction of Loch Katrine water?[53]

When Clarke described this article, he could barely contain his jubilation. He saw it as "a delicious piece of irony" and a critical piece of evidence against the claim that Loch Katrine water might prove dangerous if distributed through lead piping. Such reasoning was faulty. It was possible that, while waters from the river Clyde and the Gorbals gathering grounds were not lead solvent, water from Loch Katrine was. But even if one could have shown that waters from the Clyde and the Gorbals company were *equally* as lead solvent as water from Loch Katrine, this finding would in no way have undermined Penny's case in favor of treating water from Loch Katrine to minimize its lead solvency. Clarke probably found all of this "delicious" because it highlighted the hypocrisy of the private companies—they "had been selling lead-conveyed water to the people for half a century," but only after there was a threat of municipalization did the companies become concerned about lead poisoning. What Clarke failed to recognize, however, was that private enterprises were not the only hypocrites. The single greatest argument in favor of the Loch Katrine plan was that it promised to reduce disease rates in the city. Yet if the Loch Katrine promoters were genuinely concerned about disease, intellectual and moral consistency demanded that they show much greater sensitivity and responsiveness to the dangers of lead piping and corrosive water.[54]

Propagating a Legend

With regard to the lead question, the people who did the most to spread the legend of Loch Katrine were not the residents of Glasgow, the politicians and entrepreneurs who first promoted the Loch Katrine scheme, or even the bureaucrats who later ran the Loch Katrine waterworks. Rather, the people who perpetuated the myth the most were those who lived outside Scotland and had no connection with the Glasgow water system.

In a public health textbook published in 1901, Louis Parkes and Henry Kenwood wrote: "The Loch Katrine water acts most powerfully on lead, and yet no symptoms of lead poisoning have ever been observed amongst the population of Glasgow."[55] In a paper presented before the Institution of Civil Engineers of Ireland in December 1920, James Reade argued that "soft water supplied to Manchester and Glasgow" produced "no evil effect on the health of the consumers."[56] In an article published in *Chemical News* in 1882, a French expert wrote of the Loch Katrine water supply: "In consequence of the purity of this water, contamination with lead was much apprehended; however, the house pipes were made of this metal...and experience has shown that no inconvenience has resulted."[57]

Similarly, in his journal *The Asclepiad*, Sir Benjamin Ward Richardson maintained that Glasgow's investigation of Loch Katrine water in 1854 and 1855 clearly demonstrated the safety of the water. "Under certain circumstances," he wrote, "pure or soft waters [did] not act up lead.... This was well illustrated by the Town Council of Glasgow, in connection with the proposed water supply to that city from Loch Katrine." Richardson argued that the city's investigation cost £5,000 and had been "of the most extensive and exhaustive character." He further claimed that the investigation "proved, *inter alia*, that Loch Katrine and other equally pure or soft waters exerted, under certain circumstances, no deleterious action on lead."[58]

One of the most intriguing discussions of Loch Katrine's effect on lead came in a report issued by the British government. Published in 1894, this report came on the heels of one of the worst recorded outbreaks of water plumbism in modern history, with one source estimating that as many eight million people could have been injured.[59] The report explained that when water from Loch Katrine was removed directly from the loch and then analyzed in a laboratory it was "found to possess very vigorous and sustained action on lead." Yet this same water as delivered to Glasgow through mains and aqueducts was "found to have a very insignificant action on lead pipes, and as matter of fact lead poisoning by the water service [has never been] heard of in Glasgow."[60]

Most outside observers believed that Loch Katrine water was safe because of an insoluble coating that formed in the pipe soon after the water was introduced. In his authoritative textbook on water chemistry published in 1901, John C. Thresh wrote: "Glasgow is supplied with Loch Katrine water, which has a hardness of less than 1°, and lead service pipes are in general use; yet lead poisoning is unknown in that city." Thresh then explained that when Loch Katrine water was "passed through a pipe continuously" it coated the "inside [of the pipe] with a deposit of vegetable matter." Combining with the "oxide of lead," the vegetable matter "form[ed] a closely adherent film [that] prevent[ed] all change."[61]

Get Real

The legend of Loch Katrine is part fact and part fiction. It is difficult to disentangle fact from fiction or even to judge their relative magnitudes, because there is a dearth of empirically verifiable information about the waterworks. For whatever reason, most authors and historians appear to prefer making interpretive statements about the loch's romance, symbolism, and singular greatness.

In his epic history of the Loch Katrine aqueduct, Clarke portrayed the aqueduct as a moral and economic bargain. His evidence for this proposition was odd. He compared the amount of labor needed to build the "Great Pyramid" of Egypt with the amount required to construct the aqueduct. Citing Herodotus, Clarke maintained that it took "over one hundred thousand slaves" toiling "over twenty years" to build the pyramid, which in the end, was nothing more than "a stone cairn to protect the mummified carcass of an insignificant imbecile." In contrast, in the building of the aqueduct, "questions of profit or of vanity never once arose [because] the good of mankind inspired the promoters.... This tremendous achievement was begun and finished by less than four thousand men within the short space of three-and-a-half years."[62]

There is no question that the aqueduct was an achievement, but it should be judged against the aqueducts other modern cities built. Many large cities—including Boston (ca. 1846), New York City (ca. 1842), and Washington, D.C.—built aqueducts to distant water sources. For example, the Croton aqueduct for New York City stretched over 41 miles, had much greater capacity than the Loch Katrine aqueduct, covered difficult terrain, was built a decade earlier, and was completed in about the same amount of time as the Loch Katrine aqueduct. When judged against contemporaneous aqueduct projects in other cities, the Loch Katrine aqueduct looks like what it was: an accomplishment typical of many large cities of the era. Moreover, if a historian were so inclined, it would be relatively easy to write an ode to projects like the Croton aqueduct in New York or the Owens River Valley aqueduct to Los Angeles. The latter project covered more than 240 miles; both aqueducts were widely viewed as marvels of engineering. Yet it would seem odd to compare them to the great pyramids of Egypt, or to wax poetic

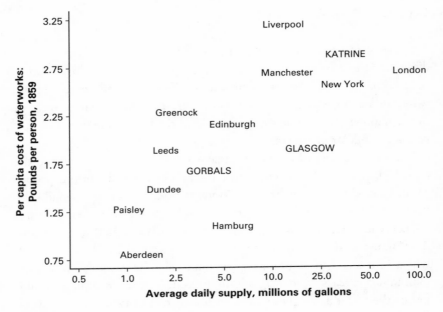

Figure 8.4
Per capita cost of various public water supplies, circa 1859. *Source:* Strang (1859).

about the "benevolent spirit of civic interventionism" they embodied. Aqueducts were built for a simple reason: voters demanded larger and less polluted water supplies, and politicians responded to this demand.[63]

One clear way to assess the costs of the Loch Katrine water system is to examine the per capita burden of the investment relative to other cities with similarly sized water systems. This is done in figure 8.4. The per capita cost of the Loch Katrine system is shown by the label "KATRINE." At the end of 1859, the total capital cost of the system, per person connected to the system, was around £3.00. For a system of its size and capacity, the Loch Katrine water system was not cheap, but rather cost what one would have expected to pay if London and New York served as guides. It is notable that of the cities plotted in the figure, only Liverpool had a higher per capita cost. Note also that the systems of the Glasgow and Gorbals water companies were relatively inexpensive, per customer, given their respective capacities. However, the Glasgow and Gorbals companies charged more for their water than did the municipally owned Loch Katrine system, and the former companies provided an inferior product.

As stated earlier, fear of disease, particularly cholera, was perhaps the single most important force behind Glasgow's efforts to municipalize its water system and introduce water from Loch Katrine. There is evidence that water from Loch Katrine improved the situation. As table 8.1 shows, the cholera epidemics that struck Glasgow were particularly severe relative to other parts of Scotland, and the disparity between Glasgow and other areas of Scotland was becoming increasingly large over time. During the epidemic that struck the city in 1853–1854, the cholera death rate in Glasgow exceeded the death rate for the rest of Scotland by a factor of 15.6.[64] The introduction of Loch Katrine water in 1860 appears to have affected a change, because during the next cholera epidemic in 1866, the death rate in Glasgow was less than half the rate that prevailed elsewhere.

As for Loch Katrine's effects on other waterborne diseases, it is noteworthy that although there were typhoid fever epidemics in Glasgow after 1860, these appear to have been traced to impure milk, not poor water. Furthermore, the death rate from diarrheal diseases, a leading killer of infants and very young children, fell sharply after the introduction of Loch Katrine water in 1860. In particular, between 1855 and 1860, the mortality rate from diarrheal diseases varied from 0.9 to 2.1 deaths per 1,000 persons, while in the five years following the introduction of Loch Katrine water, the mortality rate never rose above 0.8.[65] In his paean to municipal ownership in Glasgow, Robert Crawford described the effects of Loch Katrine water on diarrheal mortality as "magical" and "singular." But there was nothing magical or singular about Glasgow's experience with water purification. Cities all over the globe experienced sharp drops in death rates from diarrheal causes after they introduced filtered or otherwise unpolluted water supplies.[66]

Improvements in Glasgow's public water supply did not significantly affect mortality from non-waterborne diseases or overall mortality. This can be seen in figures 8.2 and 8.3, which (as noted previously) plot Glasgow's crude mortality rate and life expectancy for the period 1820–1920. Life expectancy shows a break in trend around 1840 and exhibits no clear change in trend around 1860, after completion of the aqueduct. Similarly, the crude mortality rate begins a downward trend at least ten years before the introduction of Loch Katrine water and exhibits no break in trend around 1860.[67] These patterns contrast sharply with the findings for cities like Chicago, Illinois, where improvements in the public water supply induced a radical change in the city's mortality profile. In Chicago, investments in water purification ushered in the city's transition from a high-mortality to a low-mortality environment. Loch Katrine had no such effect, perhaps because of lead contamination.[68]

Was Water from Loch Katrine Lead Solvent?

In the years following the construction of the Loch Katrine aqueduct, few people in Glasgow gave Dr. Penny and his concerns about lead poisoning much thought.[69] And of the few people who did, nearly all of them rejected the idea that the city's water supply was poisoned. Instead, they adhered to the "firmly held" belief that, while Loch Katrine water "might have [had] some slight action on new pipes," this action ceased after a short period of use "by reason of the formation of a protective coating" on the interior of the pipes. The first person to rise to the defense of Dr. Penny was the public health expert, E. J. Mills. Writing in 1891, Mills described the belief that Loch Katrine water ceased having an effect on lead pipe as "pure delusion." Mills argued that Loch Katrine water "never cease[d]" to act on lead pipes and cisterns. He argued further that "the coating in which the citizens ha[d] hitherto placed their faith [was] not insoluble, [but was] taken up by the water to an easily recognised extent."[70]

Unfortunately, few observers found Dr. Mills any more convincing than they had Dr. Penny. Less than a decade after Mills issued his statement that water from Loch Katrine was lead solvent, Benjamin Taylor published a lengthy article in the trade journal *Engineering Magazine*. After describing how water rates in Glasgow were among the lowest in the United Kingdom and explaining the complexity of the Loch Katrine aqueduct, Taylor described Glasgow's water supply as "a continuous, never failing, unrestricted stream of the purest water in the world." The evidentiary source upon which Taylor based his claim that Loch Katrine water was the "purest in the world" was never specified. Taylor did not mention the work of Dr. Mills, or the possibility that water from Loch Katrine might have been taking up unhealthy amounts of lead.[71]

Dr. Michael R. Moore was a physician and researcher at the University of Glasgow. During the 1970s and 1980s, Moore and a team of researchers published a series of articles on lead exposure from Glasgow's water supply. These articles were published in top medical and environmental journals, including the *Lancet*, the *Scottish Medical Journal*, *Clinica Chimica Acta*, and *Science of the Total Environment*. Using statistical and chemical techniques that were unavailable to his nineteenth-century predecessors, Moore established that scientists like Dr. Penny and Dr. Mills were right: Loch Katrine water was lead solvent and should have been treated with lime to limit its corrosiveness, or the use of lead service pipes should have been abandoned.[72]

None of these things happened, however—at least not until the late twentieth century, after more than one hundred years of unnecessary illness. Furthermore, the findings of Dr. Moore, while an accurate measure of the problem during the late twentieth century, understate the extent of the problem during the 1800s and early 1900s. As explained in earlier chapters, because the lead pipes were relatively new in the nineteenth century, they would have been more vulnerable to corrosion and would have been associated with higher lead levels then than they were a century later. Moore and his colleagues presented evidence that the lead in Glasgow's water supply was "associated with negative health effects, including ischaemic heart disease, renal insufficiency, gout, and hypertension."[73] These researchers also presented evidence that elevated water-lead levels were associated with mental retardation among children in the city.[74]

Glasgow began dosing Loch Katrine water with lime in 1979. Although it took more than a year to perfect this system, it greatly reduced the power of Loch Katrine water to dissolve lead from the interior of pipes. In a short communication to the *Lancet*, Moore explained that "before lime-dosing more than 50% of random daytime water samples taken within the" city of Glasgow had lead concentrations in excess of the World Health Organization (WHO) standard. Within a few months, only 20 percent of household water samples had lead concentrations exceeding the WHO standard. Subsequent improvements in the lime-dosing system drove down lead levels even further, and by August 1980, only 5 percent of the household samples exceeded the WHO standard.[75] Reducing the amount of lead in Glasgow's water supply had an immediate and beneficial effect on the blood-lead levels of city residents: Mean and median blood-lead levels fell by 51 and 60 percent, respectively.[76] By the year 2000, medical researchers in Glasgow proclaimed lead levels in the city's water supply sufficiently low that they could be considered "a relatively minor health problem."[77]

The most significant aspects of Moore's work related to water-lead levels and blood-lead levels during pregnancy. Drawing random samples from pregnant mothers residing in Glasgow, Moore showed that maternal blood-lead concentrations were positively correlated with the amount of lead in the mother's household tap water. Although this finding was challenged by two subsequent studies, Moore showed that both studies were flawed because they assumed that water lead and blood lead were related in a linear fashion. In fact, they were related non-linearly; the blood-lead/water-lead gradient increased most rapidly at low levels of exposure.[78] Also significant was the finding that the incidence of elevated blood-lead levels among pregnant mothers in Glasgow was 19 times higher in the years before lime-dosing than in the years immediately following.[79] Subsequent research suggested that undue lead in Glasgow's water supply was correlated with reduced birth weight and developmental delays.[80]

The discovery that Glasgow's water supply was contaminated with lead was not an isolated finding. A survey of British households conducted in 1975–1976 found that 7.8 percent of household water supplies in England and Wales had lead concentrations exceeding the WHO limit. In Scotland the problem was far worse: 34.4 percent of all domestic water samples had lead levels in excess of the WHO standard. In light of these findings, the Royal Commission on Environmental Pollution recommended that the government launch an aggressive, state-funded campaign to remove lead water pipes and replace them with pipes made of safer materials. However, describing the situation more than a decade after the survey, one British physician wrote that such recommendations had been "largely ignored," and publicity about the problem had been "largely nil." Furthermore, the remedial measures that had been adopted were "concentrated on reducing" lead solvency "rather than lead pipe replacement, despite surveys showing that water treatment alone was not always sufficient in reducing water lead to acceptable levels."[81]

Rethinking Municipal Socialism

When reformers in the United States and Europe championed the cause of municipal socialism in the water industry, they did so believing that public ownership would greatly reduce deaths from all sorts of water-borne diseases. The central argument of municipal socialists was that because private water companies were motivated by profit, public health concerns were taking a back seat to dividends and cost-saving measures. In contrast, because publicly owned enterprises were animated by noble concerns, such as winning reelection for local politicians and generally satisfying the will of the electorate, these companies would presumably respond more constructively to public health problems. Whatever the merit of such claims, they were widely held around 1900.[82]

More recently, events in Europe and the Americas have prompted renewed concerns regarding the use of private companies to provide public water supplies. One clear statement of such concerns, at least with regard to water-related lead poisoning, was made by Robin Russell Jones in a survey article published in the *Lancet*. Writing in 1989, amid the move to privatize water systems all over the United Kingdom, Jones argued that tightening "water quality standards for lead [will be] fiercely resisted by a privatised industry." For Jones, there was "an irreconcilable conflict of interest between providing water for profit and supplying water that is safe and wholesome to drink." Given the enormous costs of replacing old lead pipes, such concerns seem appropriate, but the case for public provision is not so clear cut.[83]

On a purely theoretical level, there is reason to believe that private companies would perform as well as or better than public companies. Consider the claim that private water companies do not have an incentive to provide clean and pure water. A company that routinely sold water tainted with typhoid fever and cholera would soon find itself with few customers as people died, fled the city, or switched to alternative water sources. Furthermore, legal and political institutions frequently impose costs on private water companies for failing to provide high-quality water, so those companies might well have stronger incentives to prevent disease than would a publicly owned company.

Recent statistical studies provide ample evidence that private water companies often provide more disease prevention than do public companies. One widely cited study finds that privatizing water systems in Argentina reduced infant mortality rates by 25 percent or more.[84] Similarly, one of the strongest forces behind the privatization of municipal water companies in the United States today is that the publicly owned systems lack the capital necessary to bring their systems into compliance with current EPA standards.[85] In early twentieth-century America, private water companies were more likely than public companies to have installed filtration systems.[86]

The standard assumption is that publicly owned water companies act in benevolent and public-spirited ways. History, however, is replete with examples that suggest otherwise. Consider the experiences of Glasgow and New York City. After 1854, the water systems in Glasgow and New York City were publicly owned and operated. By 1860, there was evidence in both cities that their water supplies were corrosive and taking up enough lead to make people sick. Yet both cities continued to use and install new lead water pipes throughout the nineteenth and early twentieth centuries. It took more than one hundred years for either Glasgow or New York to begin treating their water to minimize lead solvency, long after the most serious public health effects had passed. Whatever else might be said of the people who ran the public water systems in Glasgow and New York, they were not characterized by an excessive concern over lead poisoning. On the contrary, officials in both cities ignored, and in some cases, ridiculed, anyone who claimed that the water supplies were taking up dangerous amounts of lead.[87]

The actions of officials in New York and Glasgow contrast sharply with those of Milford Water Company. As shown in chapter 3, the Milford Water Company was a private enterprise that responded more quickly to evidence of undue water lead than did either Glasgow or New York. One might object to this comparison by arguing that New York and Glasgow were both large cities, while Milford was small. Perhaps it was more difficult and costly to prevent lead contamination in large cities than in small ones. The central problem with this line of thought is that water treatment to prevent lead solvency was subject to economies of scale, and as a result, the per capita costs of water treatment would have been lower in Glasgow and New York City. A more serious objection to the examples above is that they have not been chosen systematically, and as a result, they might not be representative of all cities and towns.

This last objection can be addressed by looking at the available data. In 1903, the U.S. Census Bureau published data on the 112 cities in the United States with populations greater than thirty thousand. By combining these data with information from the *Manual of American Waterworks*, it is possible to analyze the use of lead service pipes more systematically.[88] Of these cities, 26 had water systems owned and operated by private enterprises; 86 had water systems operated by the municipality. Use of lead service pipes was more pronounced in the cities with public water companies: 71 percent of the municipal companies employed lead service pipes, while 65 percent of the private companies employed lead. This simple comparison survives more rigorous statistical and econometric testing.[89] The conclusion of this discussion is not that private institutions necessarily outperform public ones in dealing with lead-contaminated water supplies. Rather, the point is that one should not blithely assume that public enterprises are above reproach, or necessarily superior to private organizations in the provision of public health.

Summary

The central thesis of this book is that water-related lead poisoning represents one of the world's great environmental disasters. Yet few historical observers would have ever classified it as such, and most people today are unaware that lead water pipes were widely used in the modern world, let alone constitute a source of disease. If lead-poisoned water was such a serious problem, one might ask, how come no one has ever heard of it? How could something so significant have gone unnoticed and undetected for so long? It seems impossible. Glasgow, however, shows that it is possible—the city distributed lead-poisoned water for more than a century and a half before anyone really noticed. Glasgow also shows how the apparently impossible could have happened, how the confluence of an ambitious ideology and crude science blinded policymakers to the risks posed by water distribution systems based on lead.

Municipal socialism, particularly with regard to the water industry, was predicated on the idea that private enterprise was corrupt and incapable of providing adequate levels of disease prevention. The experiences of Glasgow, New York City, and Milford challenge this assumption, demonstrating that private water companies could be more attentive to the dangers of lead contamination than publicly owned companies. A systematic comparison of lead usage across large American cities around 1900 shows that lead water pipes were no more common in cities with private water companies than in cities with public companies.

9

Building on the Past

How did this environmental disaster happen? Why did so many people—from policymakers and newspaper editors to plumbers and ordinary homeowners—believe lead water pipes were perfectly safe and adopt them en masse? Part of the answer is that the science of lead had yet to develop and, as a result, it was a struggle for observers like Alfred Swann to convince voters and policymakers of the dangers posed by lead water pipes. Most scientists during the nineteenth century appear to have used a simple heuristic: If much of the adult population in a particular city was ill with lead-related pathologies and that population was not occupationally exposed to lead, then, and only then, one could claim that lead water pipes represented a public health problem. If, however, few adults were made ill, lead pipes posed no risk. The problem with this approach was that adults were the wrong barometer. If there was enough lead in the water to induce only minor symptoms in adults, it is likely that the effects on the developing fetus were much larger (see chapters 3, 5, and 7).

More generally, three institutional and ideological forces shaped the struggle to define the role of lead in public water systems. The first was economics. Water service pipes were expensive investments. There was a reluctance to replace already installed service pipes unless there was overwhelming evidence that the local water supply was lead solvent and capable of causing significant harm to adult health. Ironically, the financial incentives to use lead pipes were strongest in those regions where lead pipes were the most dangerous. In regions with soft or otherwise corrosive water, iron and galvanized pipes would often be eaten away by the water within a few years' time. Lead pipes cost more, but were corroded much more slowly (see chapters 6, 7, and 8).

The second force shaping the use of lead water pipes was ideology. The construction of urban water systems was often tied to ideas about the appropriate role of local governments in some areas of economic activity, notably public utilities such as water, gas, and electric. The proponents of municipal socialism argued that these activities could not be left in the hands of profiteering capitalists because they could not be trusted to make public health and the broader welfare of city residents their top priority. In Glasgow, municipal socialists used the newly constructed Loch Katrine waterworks as the quintessential example of what their movement could accomplish, and this pride undermined their ability to admit that the new water system was prone to leach lead, even though the problem could have been fixed for a pittance (see chapter 8).

The third and final force was the law. Throughout the nineteenth century, courts in England and the United States refused to hold urban water providers, whether publicly or privately owned, liable for any harm that might have resulted from the use of lead water pipes. Because the courts held homeowners liable for damages, there were limited incentives for public water providers to adopt measures protecting consumers from lead exposure related to service pipes, and household plumbing more generally (see chapter 7).

Decline and Resurgence

Although policymakers and engineers were slow to acknowledge the dangers of lead water lines, over time municipalities began to abandon the practice of using lead pipes, and the lead pipes that remained were gradually coated so that less lead was taken off the interior of the pipes. By the 1930s, the federal government and some state governments were actively regulating the amount of lead that could be used in household plumbing systems. In Massachusetts, officials appear to have been even more aggressive and, as early as 1890, the State Board of Health was recommending that municipalities in the state abandon lead pipes. In neighboring New Hampshire, public health officials began publicizing the dangers of lead service pipes during the early 1900s. According to one

Table 9.1
Lead content in New Hampshire tap water, 1909–1922

Period	No. of samples	% of water samples with lead content > 0.5 ppm
1909–1910	699	49.5
1911–1912	677	39.0
1913–1914	632	45.0
1915–1916	581	33.2
1917–1918	371	24.0
1919–1920	439	21.4
1921–1922	380	18.7

Source: Howard (1923).

official, lead water pipes were abandoned "by the score" following investigations into the lead levels of New Hampshire tap water. Homeowners, however, gave up their lead pipes "always with more or less reluctance," keenly aware of the "superior merits of lead over iron, not only as to durability and ease of laying but as regards freedom from rusting, and in the tendency to deliver cleaner water."[1]

Campaigns to eliminate lead pipes appear to have had a significant effect on the lead content of household tap water. Consider again the experience of New Hampshire. Tap water was regularly sampled from around the state for lead content. Table 9.1 reports the proportion of tap water sampled that contained lead in excess of 0.5 ppm. During the two-year period of 1909 and 1910, half of the nearly 700 samples of household tap water contained lead in excess of 0.5 ppm. By 1921–1922, the proportion of samples with lead levels greater than 0.5 ppm was slightly less than 20 percent. Although by today's standards a lead level below 0.5 ppm is not especially praiseworthy, this reduction in lead levels would have had a substantial effect on infant mortality rates. The regression results reported in appendix A suggest that, in the typical city, reducing lead levels in tap water from 0.75 ppm to 0.25 ppm would have cut infant death rates by 16 percent.

Long-term trends also appear downward if for no other reason than that the exposure level considered safe by government authorities has been falling over time. In turn-of-the-century Massachusetts, health officials stated that any lead level below 0.5 parts per million was safe, while today federal officials in the United States place that threshold at 0.015 ppm. Put another way, the amount of exposure considered safe in 1900 was 33 times greater than the amount considered safe today. The historical record is replete with examples of water in the United States and England containing lead levels that were hundreds, and sometimes thousands, of times greater than the modern EPA standard. The same cannot be said of the past twenty or thirty years, even in regions such as rural Scotland and soft water areas in England, where water lead had, until recently, been a serious problem.[2] Compare, for example, the lead levels discovered in New York City tap water during the 1850s and 1860s to those found in the city in the past twenty years. The nineteenth-century levels were sometimes several hundred times the current threshold, while those from the past two decades exceeded current guidelines only occasionally, and even in those cases, the samples contained lead levels only 2 to 3 times the modern threshold.

It would be a mistake, however, to suggest that water-lead levels have fallen by similar magnitudes in all regions and urban centers. Consider the experience of Boston, Massachusetts. During the 1890s, researchers documented water-lead levels in the city of 67 to 133 times the modern EPA standard (0.015). A survey conducted by the State Board of Health in 1900 found Boston tap water usually contained lead levels between 10 and 30 times the modern EPA standard. During the 1970s, two separate studies explored lead levels in Boston tap water. Of the three hundred water samples, more than half contained no lead, and at least 70 percent had lead concentrations below the current EPA threshold. The mean lead level in city tap water was 0.03 parts per million, only 2 times the current threshold. However, one water sample contained lead in the amount of 1.5 ppm, 100 times the current EPA standard of 0.015. While this observation might have been an outlier caused by the sample having stood in the service pipe for an inordinate period of time, running water from the same home contained 0.13 parts per million, just under 10 times the current EPA standard.[3]

It would be a mistake to suggest that because water-lead levels have fallen, water lead is no longer a concern. In the late twentieth century, there remained many places with excessive water-lead levels. For example, Rio de Janeiro, the Vosgain Mountain region in France, and southern Saxonia in Germany appear to have had excessive water-lead well into the 1980s.[4] In focusing on countries in Continental Europe and Latin America, these examples are not meant to suggest that England and the United States have eliminated all important sources of water lead. In 2004, S. D. Bryant published the results of a survey of 292 schools in the Philadelphia area. This survey measured water-lead levels in the drinking fountains in each school. Bryant found that 12 percent of all schools in the area had water-lead levels at least 6.7 times the modern EPA standard, and a majority of schools (57 percent) had water lead levels in excess of the standard.[5]

Three recent environmental developments suggest that water lead should continue to be a public health concern of at least moderate importance and interest. First, increased water and air pollution are associated with increased acidity and lead solvency of public water supplies. There is evidence that acid rain causes water to take up more metals from the soil and water distribution systems, including lead, cadmium, mercury, and aluminum.[6] A recent chemical spill near Camelford, England, caused neighboring water supplies to become acidic and delivered the populations connected to those supplies a large dose of lead.[7] It is notable that physicians concerned about the lead solvency of public water supplies appear to have been among the first observers to draw attention to the dangers of the burning of fossil fuels and associated acid rain (see chapter 6).

Second, new research indicates that water can transmit hitherto unknown pathogens, or pathogens that, until recently, were not widely considered waterborne. In particular, cryptosporidium, the hepatitis A and E viruses, and many enteric viruses can be spread through tainted water.[8] Eradicating these pathogens is difficult because they are very small and often immune to traditional methods of water treatment.[9] Unfortunately, newer methods of water treatment, while able to effectively destroy these emergent organisms, also tend to cause the water to become more lead solvent. Recent episodes in Milwaukee and Washington, D.C., illustrate the trade-off between lead solvency and water purification.[10] In Milwaukee, efforts to reduce lead solvency helped give rise to a serious outbreak of cryptosporidium in the city during the 1990s.[11]

Conversely, in Washington, D.C., changes in the city's water treatment that were adopted in an effort to eradicate various emerging pathogens resulted in high concentrations of lead in the city's tap water in 2001 and 2002. There was also some concern regarding carcinogenic by-products generated by previous modes of water treatment.[12]

Finally, between 1960 and 2000, epidemiologists and medical researchers published more than thirty articles exploring the relationship between water hardness and cardiovascular disease. These studies leave little doubt that as the hardness of drinking water rises, mortality from cardiovascular disease falls. The central question surrounding this literature relates to the causal mechanisms that drive this correlation. Many researchers point out that hard water contains more calcium and magnesium than soft water, and these two elements, particularly magnesium, have cardioprotective effects. Hence, it is their hypothesis that a deficiency of calcium and magnesium in soft water areas drives the relationship between water hardness and cardiovascular disease. A smaller group of researchers attribute the correlation to the comparative ability of soft water supplies to absorb toxic metals, such as lead and arsenic, from distribution systems.[13]

Toward a New Environmental History

Over the past ten years, historical economists have devoted much attention to long-term changes in the health and welfare of the human population. One of the most ambitious projects in this regard has been Richard Steckel's efforts to systematize the study of skeletal remains and other archeological evidence. Collaborative and interdisciplinary, Steckel's work synthesizes the methods of archeologists, anthropologists, demographers, and economists to identify the absolute and relative health effects of the second agricultural revolution, the rise of organized human societies, and the Industrial Revolution.[14]

Although this book is an outgrowth of these recent developments, I hope that it will also suggest new directions for historical economists. Most of our efforts to understand the evolution of human health and welfare have been focused on biological phenomena, such as variation in the infectious disease environment and improvements in nutrition.

Whether or not one agrees with the proposition that the handmaidens of industrialization were "disruption, deprivation, disease, and death," it is clear that economic and demographic change are often associated with environmental degradation. Yet there has been no effort to assess, in a systematic manner, long-term changes in human exposure to inorganic poisons and the effects of those poisons on the broader environment.

Appendix A
Estimating the Effects of Lead Water Pipes on Infant and Fetal Mortality

To identify the effects of lead water mains on fetal and infant death rates, variants on the following generic equation are estimated using cross-sectional data from the year 1900:

$$y_i = \alpha_0 + \phi l_i + x_i \beta + e_i, \tag{A.1}$$

where y_i is the infant mortality rate, or the stillbirth rate, for city i; l_i is a dummy variable that assumes a value of 1 if city i employed lead service lines, and 0 if the city employed iron, cement, or some other non-lead pipe; x_i is a vector of control variables that measure such things as population, infectious disease environment, and the development of infrastructure related to public health; and e_i is a random error term.

There are two concerns with the regression specified in equation A.1. First, death rates might have been measured with greater accuracy in large towns than in small ones. To address this concern, all of the observations are weighted by population.[1] Second, it is possible that the use of lead water mains was correlated with some unidentified factor that also influenced the health outcomes of young children. In that case, the estimated coefficient on the dummy for lead water lines will be biased. Theoretically, it is not clear whether the estimated effects of lead would be biased upward or downward. One could imagine a world in which only cities that were relatively careless about health used lead service lines, in which case the coefficient on the lead-use dummy would be biased upward. Alternatively, one could imagine a world in which cities that already had serious public health problems avoided using lead service lines for fear of exacerbating their problems any further. In this case, only towns with relatively low death rates would have been willing to risk the dangers associated with using lead service pipes, and the coefficient on the lead dummy would be biased downward.

For the moment, let the focus be on factors that might impart an upward bias on the estimated effects of lead water lines. In this regard, consider a world in which the use of lead water mains were correlated with something one might call the "health consciousness" of a community. Simply put, communities that were very health conscious, and were attuned to the dangers of using lead pipes, installed iron or cement-lined pipes, while communities that were less health conscious and indifferent to the dangers of lead installed lead service lines. Whatever its effect on the probability of using lead water lines, health consciousness would also have had an independent and significant effect on infant mortality rates. The result would have been lower infant mortality rates in the cities using non-lead pipes than in those using lead pipes, even if both sets of cities had chosen to use the same type of pipes. Letting h_i indicate the level of health consciousness in town i, this situation can be formalized by respecifying equation (3.1) as

$$y_i = \alpha_0 + \phi^T l_i + \delta h_i + x_i \beta + e_i. \tag{A.2}$$

If one accepts the plausible assumption that h_i is negatively correlated with both y_i and l_i, it is clear that when h_i is omitted from the regression, as in equation (A.1), the estimated coefficient on lead pipes will reflect the effects of both h_i and l_i, and will be biased upward so that $\phi > \phi^T$.

To address the concern about omitted variable bias, five strategies are pursued. The first strategy explores the health effects of lead water pipes using age-specific death rates. This strategy flows from the idea that very small children would have been more vulnerable to the effects of lead than older, more developed individuals. If a developing fetus could withstand the relatively high levels of exposure in utero, he or she probably would have been able to withstand the relatively small amount of lead passed through breast-milk, or formula diluted with water. Furthermore, given that smaller children experienced more severe health effects from lead than older children, holding everything else constant, one expects that the use of lead water pipes would have had much larger effects on, say, stillbirth rates and infant death rates (children aged zero to one) than on death rates of older children, say those aged four to ten. If the use of lead lines had been correlated with health-consciousness, one would not necessarily expect to observe this pattern.

The second strategy explores the correlation between the use of lead service pipes and infectious disease rates. If it were a lack of health consciousness, and not lead water pipes per se, that drove up infant death rates in towns with lead water lines, one would expect to observe a positive correlation between the use of lead water lines and infectious diseases such as scarlet fever, measles, and the like. A positive relationship between lead use and infectious disease might also reflect the fact that exposure to unhealthy lead levels at a young age left children weak and compromised in terms of their ability to fight off infectious diseases. Hence, a positive relationship between lead use and infectious disease rates does not, by itself, prove that it was a lack of health consciousness that caused the elevated rates of infant death in towns with lead lines. However, a negative coefficient, or a very small positive one, on the lead dummy would certainly undermine the case for health consciousness.

The third strategy is to control for health consciousness through direct and remedial steps. For example, health consciousness probably would have manifested itself in other public investments. Presumably, health-conscious voters would have demanded the construction of sanitary sewers and encouraged local officials to use ground water for the public water supply, since ground water was much more likely to be germ-free than above-ground sources. It is a simple matter to include controls for such investments. By the same token, holding everything else constant, health consciousness would also have manifested itself in lower infectious disease rates. As the previous discussion suggests, health-conscious towns would have been more aggressive than those that were indifferent to health outcomes in undertaking the public and private precautions necessary to prevent the spread of infectious diseases. Again, it is a simple matter to control for such factors by adding variables that measure infectious disease rates observed in cities.

The fourth strategy is to employ instrumental variables. The goal here is to identify an instrument that is correlated with lead use, but not health outcomes. Reasonable instruments are available and these are related to the financing of service pipes, the complexity of urban infrastructure, the number of large industrial users of water, and the ownership history of the local water system. As will be made clear below, each of these variables has a solid theoretical and historical justification for use as an instrument.

The fifth strategy uses interaction effects to create exogenous variation. The sources of this variation are twofold. First, as explained in chapters 3 and 6, soft water supplies generally dissolved more lead than hard water supplies. This correlation suggests that the effects of lead water pipes would have been larger in towns with soft water supplies. The statistical analysis that follows shows that infant mortality rates (and stillbirth rates) were negatively correlated with water hardness in towns with lead water lines, but were uncorrelated in towns with non-lead water lines. It is difficult to argue that water hardness was negatively correlated with health consciousness in cities with lead lines, but somehow uncorrelated in cities with non-lead lines. Second, more lead was leached from new service pipes than from old ones. There is evidence that infant mortality rates (and stillbirth rates) were positively correlated with the proportion of water mains newly installed in towns with lead service pipes, but were uncorrelated in towns with non-lead service pipes. It is difficult to argue that the proportion of water mains newly installed was correlated with health consciousness in cities with lead water mains, but uncorrelated with health consciousness in cities with non-lead mains.

Sources, Composition, and Statistics

The empirical strategies just described were estimated using data originally compiled by the Massachusetts State Board of Health. Massachusetts collected and reported data on age-specific death rates and stillbirth rates for every town in the state with a population greater than five thousand persons. The Massachusetts mortality data are well known and have been widely used by historical demographers, largely because of their reliability, accuracy, and unusual detail.[2] The data used here come from the 1900 *Annual Report of the Massachusetts State Board of Health*. In addition to the data on health outcomes, data about water systems and the use of lead water lines across municipalities were also needed. These data come from *The Manual of American Waterworks*, a volume compiled and edited by Moses N. Baker, a prominent public health expert and the editor of the *Engineering News*, the leading trade journal for the municipal water industry. In one set of regressions (those dealing with the age of local water mains), the data from Baker are supplemented with data from the 1890 volume of the *Social Statistics of Cities* originally compiled by the United States Census Office.

Based on these sources, a sample of seventy-four Massachusetts towns was constructed. For most towns in the sample, there is information available about the following (municipal-level) characteristics in the year 1900: age-specific death rates; cause-specific death rates; population; whether the town used lead water pipes or some other material such as iron; hardness of the local water supply; source of local water supply (i.e., above ground or below ground); and proportion of water mains installed during the previous ten years. For a small subset of the full sample (about twenty towns), there is also data on the amount of lead contained in household tap water from each town.

For sample composition, summary statistics are provided in table A.1. As noted earlier, in all of the regressions, observations are weighted by population to control for the possibility that death rates were measured with greater accuracy in large towns than in small ones. Accordingly, summary statistics are provided for the weighted and unweighted samples. In the unweighted full sample, the median town had a population of 11,319, the smallest town had a population of 4,587, and the largest town had a population of 118,421. Although the sample does not include Boston—by far the largest city in Massachusetts—the distribution is still skewed, with a mean population (21,713) that is almost double the median. (Boston was dropped from the sample because it is such an outlier in terms of population.)

Thirty percent of the towns in the full sample used lead water lines exclusively or in tandem with some other type of material; the remaining 70 percent used no lead pipes whatsoever. Infant mortality rates in this sample of towns was high (349 deaths per 100,000 persons), but not unusually high for the urban Northeast.[3] The death rate from typhoid fever, an indicator of the overall quality of municipal sanitation, averaged 20 deaths per 100,000 persons in this sample; for the nation as a whole, the death rate from typhoid was 36 deaths per 100,000 persons.[4] The index of water hardness varies from about 1 to 6, with a mean value

Table A.1
Summary statistics

Variable	Unweighted sample			Weighted samples		
	Full	Lead	No lead	Full	Lead	No lead
Stillbirths per 100,000 persons	94.6 (56.5)	106.5 (76.3)	89.8 (45.7)	121.9 (76.5)	163.1 (102.7)	95.9 (36.4)
Death rate: children 0–1 years old[a]	349.4 (159.1)	368.8 (216.7)	341.2 (129.1)	425.3 (170.2)	523.6 (209.2)	363.0 (102.1)
Death rate: all persons older than age 10[b]	1151.7 (261.7)	1171.6 (252.6)	1143.7 (267.3)	1119.9 (182.8)	1108.8 (161.5)	1125.0 (194.6)
Population, in thousands, 1900	21.7 (25.1)	28.3 (29.2)	18.9 (22.9)	50.4 (37.0)	57.0 (35.0)	46.2 (38.0)
Typhoid: deaths per 100,000 persons	19.7 (17.5)	16.7 (13.5)	21.0 (19.0)	21.2 (12.9)	19.3 (9.58)	22.4 (14.6)
= 1 if lead water pipes	0.297	...	...	0.388	...	...
Hardness of water	1.85 (1.56)	1.56 (1.16)	1.97 (1.69)	1.61 (1.29)	1.40 (0.797)	1.74 (1.52)
% of all water mains < 10 years old	0.209 (0.107)	0.229 (0.137)	0.199 (0.089)	0.222 (0.105)	0.229 (0.113)	0.218 (0.101)
No. of observations	74	22	52	74	22	52

Sources: Massachusetts State Board of Health (1900), pp. 490–493. See also Massachusetts State Board of Health (1899); Baker (1897).

Note: The weighted sample gives towns with larger populations more weight than towns with small populations. This is done to control for the possibility that death rates were measured with greater accuracy in large towns than in small ones. The data are weighted using the aweights algorithm in STATA (7.0). This weighting scheme uses weights that are inversely proportional to the variance of an observation so that the variance of the jth observation is assumed to be σ^2/w_j, where w_j is the weight of the jth observation.

[a] Number of infant deaths per 100,000 persons.

[b] Number of deaths among persons aged ten or older per 100,000 persons.

of 1.85. (Higher numbers indicate harder water.) In the typical town, about 20 percent of all water mains had been installed within the previous ten years.

As for the subsamples, towns that used lead water pipes were, on average, almost 50 percent larger than towns that used non-lead pipes. Although infant mortality rates, stillbirth rates, and adult mortality rates were all higher in towns with lead pipes, typhoid rates were slightly lower. Towns with lead pipes tended to have softer water than towns with non-lead pipes; the index of water hardness was nearly 25 percent higher in towns without lead. Towns using lead and towns that did not were qualitatively similar in the proportions of their water mains that had been installed within the previous ten years. It is noteworthy that the adverse effects of lead water lines appear larger in the weighted sample than in the non-weighted sample. This is a potential concern as it might suggest that the use of lead is correlated with some unobservable variable related to both city size and health outcomes. Although concerns about unobserved heterogeneity are formally addressed through instrumental variables and other techniques, remedial statistical work suggests that lead water lines are not picking up some health effect related to city size.[5]

Age and Vulnerability to Fatal Lead Poisoning

Consider first the econometric results exploring the relationship between age and susceptibility to fatal lead poisoning. The results are reported in table A.2. Aside from measures of lead exposure, the regressions include controls for city size (log of population), and overall quality of public infrastructure (as proxied by typhoid rates). Regressions (1)–(6) use a direct measure of lead exposure: the log of the lead content in tap water from a sample of twenty-one Massachusetts towns. For these twenty-one towns, the lead content of tap water had a large and statistically significant effect on the death rates of infants and children aged one to two. For infants, an increase in lead levels by one standard deviation would have generated an 18 percent increase in mortality rates. For children aged one to two, a one-standard-deviation increase in lead would have generated a 12 percent increase in mortality rates. Lead levels have no systematic effect on the mortality rates of children aged two to four, or on children aged four to ten. Surprisingly, lead levels do have a statistically significant effect on the mortality rates of all persons older than ten, but the effect is small; a one-standard-deviation increase in lead levels only increased adult mortality by 3 percent.

Regressions (7)–(12) identify the average effect of lead water lines on the mortality rates of young children. Notice that the sample size has more than doubled, because there are data on the use of lead water lines for many more cities than there are data on the exact level of lead in tap water. The results here conform perfectly to the prediction that lead water pipes would have affected the very young disproportionately. The use of lead water lines had a large and statistically significant effect on the mortality rates of the very young, but no effect, either statistically or substantively, on the mortality rates of older children and adults. In particular, the use of lead water pipes increased stillbirth rates by 59 percent, infant mortality rates by 39 percent, and the mortality rates of children aged one to two by 24 percent.

That lead water lines had no discernible or systematic effect on the mortality rates of older children is consistent with the idea that only the youngest and least-developed children were vulnerable to fatal lead poisoning. If the use of lead water lines per se was not undermining the health of young children, but was instead correlated with some unobserved variable called health consciousness, one wonders why the large and strong correlation between lead pipes and childhood mortality ceases to exist after children reach two years of age.

Lead Water Pipes and Infectious Disease Rates

In this section, the correlation between water lead and infectious disease is identified. Four childhood diseases are considered: diphtheria; measles; scarlet fever; and whooping cough. Table A.3 reports the regression results. The first set of regressions are for a small sample of twenty-one towns for which there is data on the lead content of household tap water. For these twenty-one towns, there is weak evidence of a negative correlation between the lead content of tap water and infectious disease rates, particularly diphtheria. For the full sample of towns, there is weak

Table A.2
Age and vulnerability to fatal lead poisoning

	mean (std dev)	Effects of lead content on mortality rates						Effects of lead services on mortality rates					
		(1)	(2)	(3)	(4)	(5)	(6)	(7)	(8)	(9)	(10)	(11)	(12)
Stillbirths per 100,000 persons	94.6 (56.5)	dep. var.	...	...	...	...	...	dep. var.	...	...	...	...	...
Death rate: children 0–1 years old	349.4 (159.1)	...	dep. var.	...	...	...	...	...	dep. var.	...	...	...	...
Death rate: children 1–2 years old	72.2 (41.7)	...	...	dep. var.	...	...	...	...	...	dep. var.	...	...	...
Death rate: children 2–4 years old	53.4 (32.6)	...	...	...	dep. var.	...	...	...	...	...	dep. var.	...	...
Death rate: children 4–10 years old	61.8 (28.6)	...	...	...	...	dep. var.	...	...	...	...	...	dep. var.	...
Death rate: persons older than age 10	1151.7 (261.7)	...	...	...	...	...	dep. var.	...	...	...	...	...	dep. var.
Log (lead content)	-3.6 (1.3)	0.14 (4.5)	47.6* (14.8)	6.4* (3.9)	1.1 (2.0)	-3.0 (2.8)	28.0* (20.5)	...	...	...	...	...	...
= 1 if city used lead water lines	0.3 (0.5)	...	...	...	...	...	...	56.9* (15.9)	139.5* (33.4)	19.8* (8.9)	-0.94 (6.6)	1.8 (5.8)	-8.8 (45.6)
Log (population)	9.53 (0.9)	28.5* (6.5)	79.3* (21.1)	16.3* (5.6)	11.0* (2.8)	8.7* (4.0)	-35.5 (29.3)	27.7* (8.1)	70.6* (17.0)	13.0* (4.5)	9.2* (3.4)	6.7* (2.9)	-19.2 (23.2)

Typhoid: deaths per 100,000 persons	19.7	0.93	0.29	0.49	−0.11	0.85*	4.4*	−0.16	1.3	0.27	0.03	0.06	2.2*
	(17.5)		(2.0)	(0.54)	(0.27)	(0.38)	(2.8)	(0.6)	(1.3)	(0.32)	(0.24)	(0.21)	(1.7)
Adjusted R^2		0.500	0.621	0.402	0.446	0.234	0.109	0.271	0.356	0.150	0.061	0.037	−0.001
No. of obs.		21	21	21	21	21	21	74	74	73	73	73	73

Sources: Massachusetts State Board of Health (1900), pp. 490–493. See also Massachusetts State Board of Health (1899); Baker (1897).

Note: A constant term is included in all regressions, but is not reported. All observations have been weighted by population to control for the possibility that death rates were measured with greater accuracy in large towns than in small towns. For more details, see the notes to table A.1.

* Significant at the 10 percent level or higher (two-tailed test).

Table A.3
Lead water pipes and infectious disease rates

	Lead content					Lead water pipes				
	Diph-theria	Measles	Scarlet fever	Whoop-ing cough	ALL	Diph-theria	Measles	Scarlet fever	Whoop-ing cough	ALL
Log (lead content)	-8.85* (6.99)	1.80 (1.73)	-0.171 (1.36)	-2.43 (1.57)	-11.2* (4.82)	...	...	...	...	...
= 1 if city used lead pipes	...	...	...	...	...	-4.83 (6.99)	2.01 (3.68)	-0.922 (2.36)	-8.66* (3.21)	-12.4 (9.98)
Typhoid: deaths per 100,000 persons	0.058 (0.430)	-0.069 (0.237)	0.117 (0.185)	0.031 (0.215)	1.14* (0.660)	-0.153 (0.261)	0.014 (0.138)	0.108 (0.088)	-0.121 (0.120)	0.848* (0.374)
Log (population)	15.5* (4.48)	-0.128 (2.47)	0.170 (1.93)	-0.005 (2.24)	15.5* (6.88)	7.47* (3.56)	0.772 (1.88)	3.46* (1.20)	1.78 (1.64)	13.5* (5.08)
Adjusted R^2	0.380	-0.097	-0.051	-0.017	0.262	0.022	-0.034	0.092	0.064	0.134
No. of obs.	21	21	21	21	21	74	74	74	74	74

Sources: Massachusetts State Board of Health (1900), pp. 490–493. See also Massachusetts State Board of Health (1899); Baker (1897).

Note: A constant term is included in all regressions but is not reported. In all regressions, observations have been weighted by population. The mean death rate (deaths per 100,000 persons) for these diseases are as follows: diphtheria, 38.4; measles, 7.86; scarlet fever, 7.68; whooping cough, 11.3; all of these, 85.0.

* Significant at the 10 percent level (two-tailed test).

evidence that the use of lead water lines was negatively correlated with deaths from infectious diseases, particularly whooping cough. The weak negative correlation between infectious disease rates and lead exposure suggests that at least some of the victims of fatal lead poisoning would have eventually died from diphtheria and other infectious diseases. The absence of any systematic evidence of a positive correlation between infectious disease rates and lead undermines the idea that it was a lack of health consciousness, and not lead water mains directly, that drove up infant mortality rates.

Direct Controls for Health Consciousness

In this section, health consciousness is controlled for directly. Table A.4 reports the results for the regressions using direct controls. Regressions (1)–(3) use death rates from two childhood infectious diseases—measles and whooping cough—and age-specific death rates for older children to capture some of the effects of health consciousness. These regressions also include death rates for older-aged children as control variables. In towns with high levels of health awareness, it seems reasonable to assume that this awareness would have manifested itself in lower infectious-disease rates. Furthermore, based on the empirical work above, it is reasonable to assume that the death rates for older-aged children were correlated with deaths related to infectious diseases, but not to deaths related to lead poisoning.[6]

When these direct controls are added to the regressions, the coefficient on the lead-water-line dummy is reduced by half in the case of stillbirths, is reduced by about 20 percent in the case of infant deaths, and is unaffected in the case of deaths for children aged one to two. Nonetheless, the coefficient remains statistically significant at high levels for all three regressions, and the estimated effect of lead water lines is still quite large. In particular, regressions (1)–(3) suggest that, on average, lead water lines increased stillbirth rates by 23 percent, infant mortality rates by 31 percent, and the death rate among one- to two-year-olds by 25 percent.

Regressions (4)–(6) add the following control variables: a dummy variable indicating whether the city used an underground water source (underground sources were much less polluted than above-ground

Table A.4
Controlling for health consciousness directly

Variable	Age and infectious diseases			Plus public infrastructure		
	(1)	(2)	(3)	(4)	(5)	(6)
Stillbirths per 100,000 persons	dep. var.	...	...	dep. var.	...	...
Death rate: children 0–1 years old	...	dep. var.	...	...	dep. var.	...
Death rate: children 1–2 years old	...	...	dep. var.	...	...	dep. var.
Death rate: children 2–4 years old	0.127 (0.238)	2.35* (0.687)	0.369* (0.160)	0.186 (0.243)	2.37* (0.711)	0.378* (0.166)
Death rate: children 4–10 years old	0.256 (0.212)	−0.801 (0.609)	0.172 (0.142)	0.249 (0.215)	−0.781 (0.631)	0.195* (0.147)
Typhoid: deaths per 100,000 persons	0.367 (0.327)	1.95* (0.941)	0.285 (0.219)	0.401 (0.336)	1.85* (0.985)	0.289 (0.230)
Measles: deaths per 100,000 persons	0.574 (0.401)	1.57 (1.15)	0.962* (0.268)	0.359 (0.423)	1.54 (1.24)	0.967* (0.289)
Whooping cough: deaths per 100,000	−0.130 (0.350)	0.799 (1.01)	0.388* (0.235)	−0.067 (0.380)	0.406 (1.11)	0.283 (0.260)
Log (population)	8.29 (4.95)	34.3* (14.2)	6.97* (3.32)	3.44 (6.06)	30.9* (17.7)	8.16* (4.14)

	(1)	(2)	(3)	(4)	(5)	(6)
= 1 if underground water source	...	...	...	−11.9 (10.2)	−43.9* (30.1)	−5.48 (7.01)
= 1 if city had storm sewers	...	...	...	−19.9 (19.4)	−3.01 (56.8)	16.7 (13.3)
= 1 if city had some sanitary sewers	...	...	...	−20.0 (25.9)	48.7 (75.8)	15.4 (17.7)
= 1 if city had sanitary sewers	...	...	...	−24.0* (16.6)	−19.6 (48.6)	−0.100 (11.3)
= 1 if both sanitary and storm sewers	...	...	...	−3.8 (14.6)	−1.68 (42.8)	2.72 (9.99)
= 1 if city used lead water lines	21.7* (9.68)	112.3* (27.9)	21.2* (6.49)	21.4* (10.1)	114.7* (29.7)	21.7* (6.93)
Adjusted R^2	0.225	0.509	0.612	0.237	0.499	0.603
No. of obs.	73	73	73	73	73	73

Sources: Massachusetts State Board of Health (1900), pp. 490–493. See also Massachusetts State Board of Health (1899); Baker (1897).

Note: A constant term is included in all regressions but is not reported. All observations have been weighted by population.

* Significant at the 10 percent level or higher (one-tailed test).

sources), and a series of dummy variables indicating the development of the city's sewer system (no sewers is the omitted category). Presumably, towns that were highly health conscious would have been more likely to seek out purer water sources, and to develop more extensive sewer systems.[7] Adding controls for the development of public infrastructure does not alter the findings reported in regressions (1)–(3) in any meaningful way. Furthermore, in two of the three regressions (the infant death rate and the death rate for one- to two-year-olds), adding controls for public infrastructure reduces the adjusted-R^2, suggesting that, collectively, these controls are not correlated with death rates for young children.

An Instrumental Variables Approach

In this section, instrumental variables are used to control for unobserved heterogeneity. The estimation proceeds in two stages. In the first stage, a probit model is estimated to predict lead use in each town. In the second stage, the predicted probability of lead use in each town is used as an explanatory variable in models of early childhood mortality.[8] Three instruments are used to predict lead use. The first instrument relates to the financing of service lines. In some instances, the water company paid for the installation of service lines. In other cases, the property owner assumed the full cost, or a substantial fraction of the cost. A dummy variable indicates whether property owners had to pay, in part or in full, for the installation of a service pipe (= 1 if property owner paid; 0 if water company paid). The shorthand term for this variable will be the "owner-pays" dummy. For the sample here, in 23 percent of the towns, the water company assumed the full cost of installing the service line; in the remaining 77 percent, the property owner had to pay some or all of the cost.[9]

The theoretical justification for using the owner-pays dummy is that property owners and water companies had different time horizons and placed different values on the future benefits of a durable service pipe. These contrasting time horizons flowed from three sources. First, once a service pipe was installed, the property owner was typically responsible for maintaining and replacing the pipe if it ruptured. Second, if real-estate markets functioned well, the value of a durable service line would have been capitalized into the value of a home or rental property. Third, the rates local water companies charged for water were heavily regulated and were subject to political pressures from local voters and politicians.

As a result of the first two factors, consumers would have placed a value on the durability of service lines, and would have been willing to pay for such durability. In contrast, because the rates local water companies charged for water were heavily regulated, it would have been difficult for water companies to fully capture the future benefits of long-lived service lines, unless they charged consumers directly for the installation of those lines. Therefore, a positive correlation between the owner-pays dummy and lead use is expected. And in the raw data, this is exactly what is observed: Of the 14 towns where the water company assumed all the costs of installing service pipes, only 1 (7 percent) used lead pipes; of the 60 towns where consumers paid for the service pipe, in part or in full, 21 (35 percent) used lead pipes.

The second instrument is the number of water meters per one hundred miles of water mains.[10] This variable reflects the density and complexity of urban infrastructure. The more water meters there were, the more complex the infrastructure, and the more desirable lead service pipes would have been. Because lead was soft and pliable, it allowed plumbers to bend and twist pipes around existing fixtures. The meters-per-main variable also reflects the number of high-volume water users in a community, because meters were mainly used for large industrial customers; households typically paid for water on a fixed (zero-marginal cost) basis related to the number of taps contained in the home. The presence of large industrial consumers paying on a per-unit basis for water might have allowed water companies to cross-subsidize small consumers, including providing them with more expensive and more durable service pipes.[11] Based on this discussion, one expects meters-per-main to have been positively correlated with lead use. The raw data support this prediction: For towns using lead pipes, meters-per-main averaged 16.4, while in towns using non-lead pipes, meters-per-main averaged 11.7 (the difference is significant at the 13 percent level).

The third and final instrument relates to the ownership history of urban water systems. In turn-of-the-century Massachusetts, the ownership history of urban water systems fell into one of three categories: water companies that had always been public; companies that had always been private; or companies that were currently public but were once private. The companies of primary interest are those that experienced a change in ownership regime from private to public—companies that had been "municipalized." Recent research shows that private water companies were generally municipalized, not because of concerns about public health or the exercise of monopoly power, but because they were a valuable source of revenue and political employment for local politicians. The private water companies most at risk for such municipalization anticipated expropriation, and adjusted their fixed investments accordingly.[12]

Following this evidence, if a private water company anticipated future expropriation of its system, holding everything else constant, it would have refrained from installing pipes that were made of expensive and highly durable materials like lead, and would have chosen instead to install cheaper, less durable pipes made of untreated iron or steel. If one uses the municipalization dummy as a measure of the ex ante risk of subsequent expropriation (it is a perfect indicator of the risk ex post), a negative correlation between the municipalization dummy and lead use is expected.

Patterns in the raw data are consistent with this line of thought. Of the forty-four water companies in the sample that had *always* been publicly owned, 36 percent used lead service pipes; of the fourteen companies that had once been private but were made public sometime before 1900, 21 percent used lead service pipes (the difference is significant at the 15 percent level).

Table A.5 reports the results of the analysis employing instrumental variables. The first two regressions are for the first-stage probits. The marginal effect of each variable (rather than the formal coefficient) is reported. In the basic model, the only exogenous variables other than the instruments are typhoid rates and log of the population. In the full model, all of the exogenous variables reported in table A.5, including age-specific death rates, cause-specific death rates, log of population, water source, and the sewer dummies, are included along with the instruments. In the first-stage probits, the owner-pays dummy and the municipalization dummy are highly significant and have a large independent effect on the probability of lead use: Having the property owner pay for the service line increased the probability of lead use by 30 to 43 percentage points, and a water system that had been municipalized was 24 to 34 percentage points less likely to have lead service pipes. The basic model predicts that 33 percent of the sample would have used lead, while the full model predicts that 19 percent of the sample would have used lead. With observed lead use around 36 percent, the basic model predicts better than does the full model.

Using instrumental variables weakens some results, but strengthens others. In particular, the coefficient on predicted lead use is not a significant correlate with the death rate for children aged one to two. However, the coefficient on predicted lead use is a significant correlate with the stillbirth rate and the infant mortality rate, and it is roughly twice the size of the coefficient on observed lead use in the comparable regression presented in tables A.2, A.3, and A.4. According to the estimates here, lead water pipes increased the infant mortality rate by 227 to 289 deaths, or between 63 and 80 percent. As for stillbirths, lead pipes increased the stillbirth rate by 42 to 127 fetal deaths, or between 44 and 133 percent. These patterns suggest that if there was some sort of unobserved heterogeneity, it imparts a downward bias in the OLS estimates.

Interaction Effects and Natural Experiments

This section exploits two conditions that induce exogenous variation in the efficacy of lead water pipes. First, the amount of lead leached into water by lead service lines depended on the hardness of the water supply. In towns using lead pipes with soft water, large amounts of lead were absorbed into the water supply; in towns with hard water, relatively small amounts of lead were absorbed. See chapter 6 and appendix C for evidence linking water hardness with lead solvency. Second, the amount of lead leached from the interior walls of service pipes varied inversely with the ages of the pipes. Holding the corrosiveness of water supplies constant, less lead was dissolved from old pipes than from new ones. See chapter 3 for evidence showing that age of pipes and water lead are negatively correlated.

Table A.5
The estimated effects of lead pipes using instrumental variables

Variable	First-stage probits		Second-stage basic model			Second-stage full model		
	Basic	Full	Still's	Infant	Age1–2	Still's	Infant	Age1–2
= 1 if consumer pays for service line	0.433* (0.087)	0.300* (0.092)	...	...	...	...	...	...
Water meters per 100 main miles	0.003 (0.004)	−0.002 (0.005)	...	...	...	...	...	...
= 1 if water company always public	omitted	omitted	...	...	...	...	...	...
= 1 if water company always private	−0.021 (0.070)	0.002 (0.254)	...	...	...	...	...	...
= 1 if water company municipalized	−0.337* (0.005)	−0.239* (0.114)	...	...	...	...	...	...
Basic model variables[a]	yes	yes	yes	yes	yes	yes	yes	yes
Full model variables[b]	no	yes	no	no	no	yes	yes	yes
Predicted lead use	0.331	0.185	127.5* (30.1)	288.5* (63.9)	13.7 (18.6)	41.9* (22.0)	227.3* (66.5)	16.8 (16.2)
Observed lead use	0.388	0.345	...	...	...	...	...	...
Pseudo-R^2	0.259	0.367	...	...	...	...	...	...
Adjusted-R^2	...	...	0.315	0.378	0.360	0.225	0.477	0.546
No. of obs.	74	73	74	74	74	73	73	73

Sources: Massachusetts State Board of Health (1900), pp. 490–493. See also Massachusetts State Board of Health (1899); Baker (1897).

Note: A constant term is included in all regressions, but is not reported. All observations have been weighted by population.

[a,b] The basic (full) model includes the typhoid rate and the log of population (age-specific death rates children for two to four year olds, four to ten years old, and over ten; death rates for measles and whooping cough; sewer dummies; and a ground water dummy).

* Significant at the 10 percent level or higher (two-tailed test).

The empirical work in this section builds on these two correlations (i.e., the tendency for soft water to dissolve more lead than hard water; and the tendency for more lead to be leached from new pipes than from old ones). The correlations suggest that, holding everything else constant, lead water lines would have had a larger effect on infant mortality rates in towns with soft water and new lead pipes than in towns with hard water and old lead pipes. Evidence that such interaction effects influenced infant mortality would be difficult to reconcile with stories about health consciousness and unobserved heterogeneity. For example, one might be able to construct an argument that hard water alone was correlated with both health consciousness and infant mortality. But to suggest that not only was there this correlation, but that it somehow differed across cities with lead water lines and non-lead lines, would require an implausible argument.

The results of these experiments are reported in table A.6. In addition to the variables reported in the table, the regressions include the following control variables: the death rate for children aged two to four; the death rate for children aged four to ten; the death rate from typhoid fever; the log of the town's population; and a dummy variable indicating whether the town drew its water from an underground source. The results suggest that hard water dissolved much less lead from the interior walls of lead pipes than did soft water, which is exactly what one would expect. Interacting the lead dummy with the measure of water hardness indicates that, holding everything else constant, towns with hard water and lead pipes had significantly lower infant and young child mortality rates than did towns with soft water and lead pipes.

For towns with many newly installed water lines (defined as the proportion of water mains installed during the previous ten years), the results are again consistent with the predictions suggested by chemistry. Towns with many new lead water lines had significantly higher infant mortality rates than did towns with few new lines. Indeed, once one enters the interaction effect, the dummy on lead lines alone is small and insignificant, suggesting that all of the effects of lead water lines on infant mortality are to be found in cities with relatively new water lines. The final three regressions in table A.6 allow for a more complex interaction effect where the proportion of newly installed mains has a larger effect in cities with populations greater than sixty thousand than in smaller cities with new lead lines. Allowing for this more complex interaction effect strengthens the results.[13]

There are two concerns surrounding this analysis of interaction effects. First, based on the raw regressions alone, it is difficult to assess the magnitudes. Second, whenever one starts interacting variables with a small data set, degrees of freedom are quickly lost and it becomes possible that the results are driven by only one or two observations. To address both of these concerns, the models specified previously are made as parsimonious as possible, and the results are then plotted graphically. Aside from lead use and the hardness of local water supplies, the only control variable will be population.

Consider first, then, how the effects of lead water pipes varied with the hardness of a town's water supply. After restricting the sample to towns with populations greater than twelve thousand persons, the infant mortality rate (y_i) in each town is regressed against a measure of the hardness of the town's water supply. Running separate regressions for towns with lead lines and towns with non-lead lines, the results are as follows:

Non-Lead: $y_i = 397.8 + \text{hardness}^* \ (-20.4)$
t-statistic (12.6) (1.70)
$R^2 = 0.138$; No. of obs. $= 20$

Lead: $y_i = 839.3 + \text{hardness}^*(-190.5)$
t-statistic (7.38) (2.90)
$R^2 = 0.512$; No. of obs. $= 10$

Figures A.1 and A.2 plot the data associated with these regressions and the estimated trend lines. The y and x axes are scaled identically in both figures. Clearly, the relatively strong correlation between the hardness of local water supplies and infant mortality rates in cities with lead water lines is not driven by one or two observations. And in terms of magnitude, variation in hardness explains a large change in infant mortality rates—from a high of around 800 deaths per one hundred thousand persons for towns with soft water supplies to a low of 250 deaths per one hundred thousand for towns with hard water supplies.[14]

Consider, next, how the effects of lead water pipes varied with the age of local water lines. After restricting the sample to towns with populations

Table A.6
Soft water, new pipes, and infant mortality: A look at interaction effects

Variable	Hard water			New water pipes			New water pipes and large cities		
	Still's	Infant	Age1–2	Still's	Infant	Age1–2	Still's	Infant	Age1–2
Controls[a]	yes	yes	yes	yes	yes	yes	yes	yes	yes
= 1 if city used lead pipes	40.8* (19.3)	223.7* (55.4)	40.9* (14.7)	19.2 (20.9)	13.5 (61.0)	−0.255 (16.1)	−7.32 (21.4)	−131.8* (48.3)	−18.0 (16.9)
Hardness of water	−2.20 (3.99)	4.97 (11.5)	−0.995 (3.05)	...	...	...	...	...	...
Hardness* (lead dummy)	−10.5 (10.9)	−63.3* (31.3)	−12.1* (8.34)	...	...	...	...	...	...
% of water mains newly installed	...	...	...	−97.7* (53.1)	−80.9 (154.7)	−15.9 (40.9)	−72.0* (50.2)	−3.80 (113.5)	−9.47 (39.7)
(% new) * (lead dummy)	...	...	...	40.7 (81.0)	450.8* (236.0)	96.2* (62.5)	41.0 (76.7)	576.0* (173.4)	117.3* (60.6)
= 1 if population > 60,000	...	...	...	...	...	...	−35.0* (17.0)	−29.6 (38.4)	3.95 (13.4)
(% new) * (lead) * (pop > 60,000)	...	...	...	...	...	...	292.9* (89.3)	1367.6* (201.8)	156.3* (70.6)
Adjusted R^2	0.230	0.547	0.524	0.256	0.532	0.518	0.354	0.755	0.560
No. of obs.	64	64	64	68	68	68	68	68	68

Sources: Massachusetts State Board of Health (1900), pp. 490–493. See also Massachusetts State Board of Health (1899); Baker (1897).

Note: A constant term is included in all regressions but is not reported. All observations have been weighted by population.

[a] The control variables include log of population, the typhoid rate, death rate for children aged two to four, death rate for children aged four to ten, and a dummy variable equal to one if the city employed ground (as opposed to surface) water.

* Significant at the 10 percent level or higher (one-tailed test).

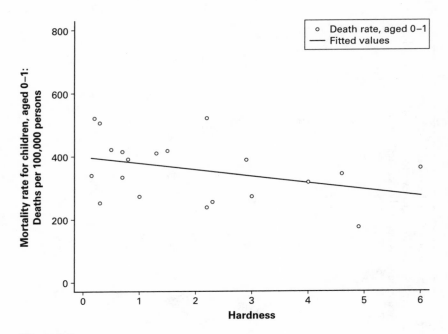

Figure A.1
Water hardness and infant mortality: Towns with no lead pipes. *Source:* Massachusetts State Board of Health (1900), pp. 490–493. See also Massachusetts State Board of Health (1899) and Baker (1897).

less than sixty thousand persons, the infant mortality rate (y_i) in each town is regressed against the proportion of the town's water mains that had been installed during the previous ten years (% new). Running separate regressions for towns with lead and towns with non-lead lines, the results are as follows:

Non-Lead: $y_i = 357.4 + (\% \text{ new})*(-68.2)$
t-statistic   (6.46)   (0.26)
$R^2 = 0.002$; No. of obs. = 42

Lead: $y_i = 95.1 + (\% \text{ new})* (886.5)$
t-statistic   (1.72)   (4.47)
$R^2 = 0.555$; No. of obs. = 18

Figures A.3 and A.4 plot the data associated with these regressions and the estimated trend lines. The y axis is scaled identically in both fig-

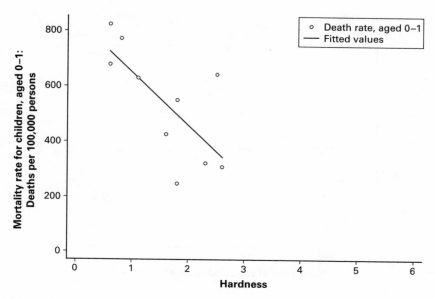

Figure A.2
Water hardness and infant mortality: Towns with lead pipes. *Sources:* Massachusetts State Board of Health (1900), pp. 490–493. See also Massachusetts State Board of Health (1899); Baker (1897).

ures. Clearly, the relatively strong correlation between (% new) and infant mortality rates in cities with lead water lines is not driven by one or two observations. And in terms of magnitude, variation in % new explains a large change in infant mortality rates—from a low of 150 deaths per one hundred thousand for cities with fewer than 10 percent newly installed mains, to a high of 750 deaths per one hundred thousand for cities with 80 percent newly installed mains.

Infant Mortality and Lead-Solvent Water in England

In this section, data from England are analyzed. Although the data are not as comprehensive as those for Massachusetts, they make it possible to replicate many of the same tests. Two data sources are employed. The first source is the *Forty-Sixth Annual Report of the Registrar-General*, which provides data on total births, illegitimate births, total

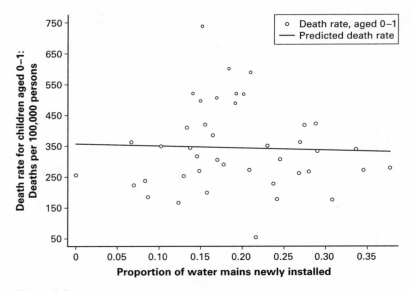

Figure A.3
New mains and infant mortality: Towns with no lead pipes. *Sources:* Massachusetts State Board of Health (1900), pp. 490–493. See also Massachusetts State Board of Health (1899); Baker (1897).

population, land area, and age-specific deaths for English towns in 1883.[15] The second source is the *Eighteenth Annual Report of the Local Government Board*, 1888–1889.[16] A supplement to this report contains an appendix identifying the towns in Yorkshire and other northern counties which used lead service pipes during the 1880s. Using these sources, data on water-related lead exposure and infant mortality have been collected for sixty-one towns in northern and midland counties.

Descriptive statistics for these data are reported in table A.7. There are two subsamples reported, "lead" towns and "no lead" towns. A town is classified as having had lead in its water if it used lead service pipes and was located in a county with soft (corrosive) water supplies; if the town did not use lead pipes, or was located in a county with hard (noncorrosive) water supplies, it was classified as having no lead in its water. Lead towns had an average infant mortality rate that was 20 percent higher than towns with no lead. Table A.8 reports regression results using age-specific death rates.

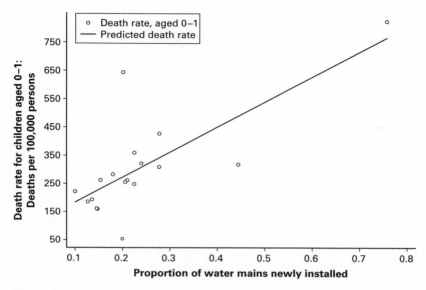

Figure A.4
New mains and infant mortality: Towns with lead pipes. *Sources:* Massachusetts State Board of Health (1900), pp. 490–493. See also Massachusetts State Board of Health (1899); Baker (1897).

As in the case with Massachusetts, one expects that lead water pipes would have affected the very young disproportionately and had little, if any, effect on the mortality rates of older children. Consistent with this prediction, infant mortality rates are 9 percent higher in lead towns than in no-lead towns (after controlling for population density, the percentage of births that were illegitimate, and the death rate for all individuals five years and older—see regression [1]). There is no evidence of any effect on the mortality rates of older children (see regressions [2]–[5]). Finally, if one uses the death rates of older children as controls, as was done in Massachusetts, the estimated effect of lead remains around 9 percent (see regression [6]).

Although the results for Massachusetts and England are broadly consistent, there is a large difference in the size of the estimated effect of lead: 9 percent for England versus 25–50 percent for Massachusetts. What accounts for this difference? There are three possibilities. First, the lead dummy in England was measured with more error than was the lead

Table A.7
Descriptive statistics, sixty-one English boroughs in 1883

Variable	Full sample Mean (Std. dev.)	Lead Mean (Std. dev.)	No lead Mean (Std. dev.)
Infant deaths per 1,000 births	135.1 (31.9)	146.5 (30.2)	122.5 (29.5)
Deaths, ages 1–2, per 1,000 persons	37.0 (18.5)	41.9 (19.3)	31.5 (16.2)
Deaths, ages 2–3, per 1,000 persons	16.1 (8.2)	17.5 (8.7)	14.6 (7.4)
Deaths, ages 3–4, per 1,000 persons	11.3 (5.6)	12.4 (5.2)	10.1 (5.9)
Deaths, ages 4–5, per 1,000 persons	8.9 (4.5)	9.6 (4.0)	8.1 (4.99)
Log (population density)	0.328 (1.79)	0.808 (1.59)	−0.201 (1.88)
(Illegitimate-births)/(total births)	0.055 (0.019)	0.054 (0.020)	0.056 (0.018)
Deaths, ages 5 and up, per 1,000 persons	12.7 (2.0)	13.6 (2.0)	12.1 (1.75)
No. of obs.	61	32	29

Sources: Massachusetts State Board of Health (1900), pp. 490–493. See also Massachusetts State Board of Health (1899); Baker (1897).
Note: These statistics are for unweighted data. Also, population density is measured as persons per acre of land in the borough.

dummy in Massachusetts. There is no way to ascertain the frequency with which lead water pipes were used in hard water counties, because government surveys did not report such information. As a result, it is necessary to assume that, even though they were common in hard water areas, lead pipes were not a significant source of lead exposure. This creates an error-in-variables problem for the England data, and suggests that the coefficient on the lead dummy is biased toward zero. However, if one restricts the sample of English towns to only those in soft water regions, for which there are accurate data on the use of lead service pipes by town, the results are unchanged: Lead service pipes increased infant mortality by around 9 percent.

Table A.8

Lead and infant mortality in sixty-one English boroughs, 1883

Variable	(1)	(2)	(3)	(4)	(5)	(6)
Infant deaths per 1,000 live births	dep. var.	—	—	—	—	dep. var.
Death rate: children 1–2 years old	—	dep. var.	—	—	—	0.010 (0.113)
Death rate: children 2–3 years old	—	—	dep. var.	—	—	0.927* (0.404)
Death rate: children 3–4 years old	—	—	—	dep. var.	—	−1.94* (0.845)
Death rate: children 4–5 years old	—	—	—	—	dep. var.	1.28 (0.732)
= 1 if lead pipes & solvent water	9.53* (4.17)	0.319 (5.59)	−0.686 (1.66)	−0.718 (0.848)	−0.761 (0.899)	9.75* (3.98)
Log (density)	11.6* (1.69)	6.40* (2.66)	4.29* (0.672)	1.46* (0.344)	1.21* (0.364)	8.90* (2.13)
% illegitimate	339.9* (141.3)	243.0 (189.3)	262.4* (56.1)	−6.06 (28.7)	39.8 (30.4)	31.3 (167.8)
Death rate: 5 years and older	2.68* (1.14)	−1.25 (1.52)	0.564 (0.452)	1.04* (0.231)	0.942* (0.245)	2.98 (1.33)
Adjusted R^2	0.695	0.092	0.609	0.656	0.543	0.728
No. of obs.	61	61	61	61	61	61

Sources: Registrar General (1883); Local Government Board (1888–1889), pp. 379–452.

Note: To control for the possibility that death rates are measured with greater accuracy in large boroughs than in small ones, all regressions have been weighted. Specifically, regressions (1) and (6) are weighted by the number of births; and regressions (2) through (5) are weighted by total population.

* Significant at the 5 percent level or higher (two-tailed test).

Second, as demonstrated by the results interacting lead pipes with water hardness and new water mains, much of the impact observed in Massachusetts was driven by a handful of cities on the upper end of the distribution: Cities with very soft and corrosive water and/or cities with a high proportion of newly installed water mains had unusually high infant mortality rates. Perhaps the sample of cities from England included no cities with similarly corrosive water supplies and new lead piping.

Third, different denominators were used for England and for Massachusetts to normalize infant deaths. Specifically, infant mortality in Massachusetts was calculated as infant deaths per 100,000 persons, while infant mortality in England was calculated as deaths per 1,000 live births. It is not clear which normalization procedure is preferable. Lead could have induced changes in fertility by making it more difficult for women to conceive, or it could have been associated with shorter spacing between pregnancies because of widespread premature births. In either case, one might argue that normalizing by the general population (as was done in Massachusetts) would be preferable to normalizing by births. If, however, leaded water had minimal effects on fertility, normalizing by births seems more appropriate.[17]

Table A.9 reports a series of regressions that help resolve these issues. In the first regression, the infant mortality rate in England is recalculated using the same normalization procedure as was used for Massachusetts: Infant mortality is now measured as infant deaths per one hundred thousand persons. Reestimating the statistical models for England with this new measure of infant mortality yields results that are larger than those reported above. The coefficient on lead now implies that water-related lead exposure increased infant mortality by 10.9 percent. By contrast, the same regression equation estimated with a birth-normalized infant mortality rate implied an increase of 7.8 percent (see table A.9, regressions [1] and [2]).

It is also possible to explore the effects of alternative normalization procedures on the Massachusetts data. The *Massachusetts Vital Statistics* of 1915 reports birth rates for all towns in that year. Assuming that birth rates were roughly similar in each town in 1900 and 1915, it is a straightforward exercise to estimate the number of births in 1900. With these estimates in hand, infant mortality rates in 1900 can be normalized

Table A.9
Alternative normalization procedures and estimated effects

Variable	England			Massachusetts		
	Mean, no lead	(1)	(2)	Mean, no lead	(3)	(4)
Birth-normalized IMR	122.5	—	dep. var.	147.6	—	dep. var.
Pop.-normalized IMR	421.6	dep. var.	—	363.0	dep. var.	—
Lead dummy	—	9.53* (4.17)	46.1* (20.7)	—	120.1* (26.1)	27.2* (8.02)
Implied change in IMR	—	0.078	0.109	—	0.331	0.184
Control variables, England	—	yes	yes	—	—	—
Control variables, Massachusetts	—	—	—	—	yes	yes
Adjusted-R^2	—	0.695	0.609	—	0.520	0.384
No. of observations	—	61	61	—	72	72

Sources: Massachusetts State Board of Health (1900), pp. 490–493; Registrar General (1883); Local Government Board (1888–1889), pp. 379–452; Baker (1897).

Note: Regressions (1) and (3) are weighted by total births in each town; regressions (2) and (4) are weighted by total population in each town. The control variables for the England regressions are the log of population density, the proportion of births that were illegitimate, a constant term, and the death rate for all persons five years of age and older. The control variables of the Massachusetts regressions are the death rate from typhoid fever, a dummy variable indicating use of a ground (versus surface) water supply, the log of population in 1900, the death rate for children aged two to four, a constant term, and the death rate for children aged four to ten. "Mean, no lead" refers to the mean IMR in towns without lead exposure (i.e., the lead dummy assumes a value of zero).
* Significant at the 0.05 level or higher (two-tailed test).

by births rather than total population; one can then compare the results for Massachusetts across the competing normalization procedures. The coefficient on lead for the birth-normalized regression implies that lead water pipes increased infant mortality by 18 percent; the same regression using population-normalized infant mortality implies an increase of 33 percent, almost double the increase implied by the birth-normalized regression. Compare regressions (3) and (4) in table A.9. Overall, the results suggest that differences in normalization can explain some of the observed difference between England and Massachusetts, but not all of it.

Appendix B

A Statistical Supplement to *The Menace and Geography of Eclampsia*

The *Annual Reports of the Registrar General* provide detailed statistics on the causes of maternal mortality across all administrative counties in England and Wales. These reports also provide information enabling one to control, at least indirectly, for industrialization, urbanization, overall health, and prenatal care. The analysis to follow reports results using data from the 1883 *Annual Report of the Register General*. Descriptive statistics for the resulting data set are provided in table B.1.

Porritt argued that industrialization and urbanization were unrelated to eclampsia. Regressing eclampsia against population density—which was highly correlated with both urbanization and manufacturing activity—corroborates this view (see table B.2, regression [1]). Porritt also argued that prenatal care and overall health, while perhaps important, were not the forces driving the geographic variation in eclampsia across England and Wales. The crude mortality rate (measured as deaths from all causes per 1,000 persons) serves as a proxy for the county's overall health status, and the illegitimacy rate (measured as the percentage of all live births born to unmarried mothers) serves as an indicator of prenatal care. Assuming that single mothers had fewer economic resources and less familial support than married mothers, it seems reasonable to argue that children born to married mothers received better prenatal care. Regressing eclampsia against the crude death rate and the illegitimacy rate suggests eclampsia was uncorrelated with either of these (see table B.2, regressions [2] and [4]).

A central issue raised by the discussion in chapter 4 is the extent to which eclampsia rates were correlated with causes of maternal mortality.

Table B.1
Descriptive statistics for eclampsia and lead, 1883

| Variable | Administrative counties with | | | |
| | Lead-solvent water | | Non-solvent water | |
	Mean	Std. Dev.	Mean	Std. Dev.
Eclampsia death rate[a]	0.561	0.277	0.235	0.171
Sepsis death rate[b]	3.243	0.827	2.255	0.693
Log (population density)[c]	−0.861	0.763	−0.420	1.400
Crude death rate[d]	23.230	3.382	20.092	3.130
Hemorrhage death rate[e]	0.868	0.327	0.840	0.352
% of births illegitimate[f]	5.572	0.009	5.572	0.012
Birthrate[g]	39.459	7.100	36.012	6.492
No. of obs.	15		31	

Sources: Porritt (1934); Registrar General (1883).
[a] Includes all deaths in childbirth caused by eclampsia during 1883. Rate is measured as deaths per 1,000 live births.
[b] Includes all deaths in childbirth caused by sepsis (puerperal fever) during 1883. Rate is measured as deaths per 1,000 live births.
[c] Population density is calculated by dividing the county's total population by acres of land in the county.
[d] Total deaths in the county per 1,000 persons.
[e] Includes all deaths in childbirth caused by hemorrhage during 1883. Rate is measured as deaths per 1,000 live births.
[f] The percentage of all births in the county that were classified as illegitimate during 1883.
[g] The birthrate is calculated as births per 1,000 persons in the county.
[h] The following counties have a history of lead-solvent water: Cheshire, Cornwall, Cumberland, Derbyshire, Durham, East Riding, Lancashire, Lincolnshire, Monmouthshire, North Riding, North Wales, Northumberland, Nottinghamshire, South Wales, West Riding, Westmorland.
[i] The following counties have a history of non-solvent water: Bedforshire, Berkshire, Buckinghamshire, Cambridgeshire, Devonshire, Dorsetshire, Essex, Gloucestershire, Hampshire, Herefordshire, Huntingdonshire, Kent, Leicestershire, Middlesex, Norfolk, Northamptonshire, Oxfordshire, Rutlandshire, Shropshire, Somersetshire, Staffordshire, Suffolk, Surrey, Sussex, Warwickshire, and Worcestershire.

Porritt suggested that there was a positive correlation between sepsis and eclampsia. Regressing the eclampsia rate against puerperal sepsis suggests a weak but positive correlation between the two variables. Regressing the eclampsia rate against the death rate from pregnancy-related hemorrhages indicates no correlation between these two variables, however (see table B.2, regressions [3] and [5]).

The final two columns of table B.2 report results from regressions of eclampsia rates against the birthrate, log of population density, crude death rate, hemorrhage death rate, percentage of illegitimate births, sepsis death rate, a dummy for lead-solvent water, and a constant. The first regression is weighted by births in each county to control for the possibility that eclampsia deaths were measured with greater precision the more births there were; the second regression is unweighted. The only explanatory variable, other than lead-solvent water, to enter the regressions significantly is the crude death rate, which has a negative coefficient in the weighted regression. The negative coefficient on the crude death rate suggests that in high-mortality counties the women most susceptible to eclampsia were dying of other causes before they ever had the chance to develop the disease.

In contrast to the other regressors, the coefficient on the lead dummy is significant and large in both regressions. The weighted regression suggests that lead-solvent water tripled the mortality rate from eclampsia; the unweighted regression suggests that it doubled the mortality rate. To see this, note that according to table B.1, counties without lead-solvent water had an average eclampsia rate of 0.235, while the estimated regression coefficients are 0.443 and 0.289 for the weighted and unweighted regressions, respectively.

The results in table B.2 suggest that lead-solvent water had an even larger effect on eclampsia rates than had been implied by Porritt's original analysis. What explains the discrepancy? One possibility is that the effects of lead-solvent water were larger in 1883 than during the 1920s. In 1883, water companies and public health officials were barely cognizant of the dangers posed by lead-solvent water. As a consequence, lead pipes were widely used in areas with corrosive water, and no steps were taken to treat the water to limit its corrosive effects. Local governments in England and Wales did not begin treating water with lime and chalk,

Table B.2
The correlates of eclampsia, 1883

| | Dependent variable: Eclampsia death rate | | | | | | |
| | Partial correlations | | | | | Full model | |
Variable[a]	(1)	(2)	(3)	(4)	(5)	Wtd.[c]	OLS[c]
Birthrate	−0.018 (0.032)	...	...	...	...	0.013 (0.011)	−0.015 (0.013)
Log (population density)	...	...	...	...	...	0.026 (0.029)	−0.002 (0.043)
Crude death rate	...	0.016 (0.011)	...	...	...	−0.057[b] (0.026)	0.017 (0.032)
Hemorrhage death rate	...	...	0.089 (0.114)	...	...	0.023 (0.128)	0.049 (0.112)
% illegitimate	...	...	...	0.319 (0.326)	...	−0.100 (0.402)	−0.311 (0.385)
Sepsis death rate	...	...	...	...	0.096[b] (0.043)	0.015 (0.054)	0.029 (0.052)
= 1 if lead-solvent water	...	...	...	...	...	0.464[b] (0.111)	0.312[b] (0.113)
Constant	0.331[b] (0.042)	0.012 (0.233)	0.265 (0.104)	0.177 (0.172)	0.096 (0.116)	0.937[b] (0.389)	0.482 (0.403)
Adjusted-R^2	−0.015	0.023	−0.009	−0.001	0.082	0.443	0.289
No. of obs.	46	46	46	46	46	46	46

Sources: Porritt (1934); Registrar General (1883).

[a] For precise definitions of the variables, see notes a–i, table B.1.

[b] Significant at the 5 percent level or higher (one-tailed test).

[c] The weighted regression is weighted by births in the county using the aweights algorithm in STATA.

which usually limited the amount of lead the water absorbed, until the 1890s.[1]

To test the relationship between lead-solvent water and the death rate from convulsions of unknown etiology, data from the Registrar General's *Annual Report* for the year 1883 are used once again. With these data, variants on the following regression model are estimated:

$$y_i = \beta_0 + \beta_1 Z_i + \beta_2 W_i + \beta_3 P_i + \beta_4 L_i + \mu_i, \tag{B.1}$$

where, y_i is the death rate from convulsions, seizures, apoplexy, and related disorders, in county i;[2] Z_i is the death rate from a class of infectious diseases known as zymotic diseases (e.g., typhoid fever and influenza);[3] W_i is the death rate from epilepsy; P_i is the log of the county's population density, measured as persons per acre; L_i is a measure of the lead levels in the county's water; and μ_i is an error term. Death rates are measured as deaths per 1,000 persons. Including the death rate from zymotic diseases controls for the overall health of the county; including the death rate from epilepsy (which was, and is, unrelated to lead exposure) controls for the possibility that some deaths from convulsions were actually cases of epilepsy, mistakenly diagnosed. The measures of lead are either the eclampsia rate (an indirect measure) or a dummy variable indicating whether lead-solvent water was widespread in the county.

The results are reported in table B.3. There are two sets of regressions. One set is unweighted; the other set is weighted by population to control for the possibility that death rates were measured with greater accuracy in large counties than in small ones. The results indicate that the eclampsia death rate was positively correlated with the death rate from convulsions of unknown etiology (see regressions [1] and [5]). There is also evidence that lead-solvent water increased the death rate from convulsions. The dummy on lead-solvent water is significant, and depending upon specification, equals 0.25 or 0.34. This suggests that, holding everything else constant, lead-solvent water increased the death rate from convulsions by 10–15 percent (see regressions [2] and [6]).

Furthermore, once the lead dummy is added to the regressions, the evidence that eclampsia is correlated with convulsions becomes much less strong, or vanishes entirely (see regressions [3] and [7]). This suggests that it was lead, not some unidentified third variable correlated with

Table B.3
Convulsions and lead-solvent water, 1883

| | Dependent variable: Convulsions death rate (mean = 2.36) | | | | | | | |
| | Unweighted regressions | | | | Weighted regressions[c] | | | |
Variable[a]	(1)	(2)	(3)	(4)	(5)	(6)	(7)	(8)
Eclampsia death rate	0.353[b] (0.156)	...	0.210 (0.172)	...	0.313[b] (0.171)	...	0.015 (0.172)	...
= 1 if lead-solvent water	...	0.245[b] (0.093)	0.184 (0.105)	−0.168 (0.307)	...	0.335[b] (0.081)	0.331[b] (0.094)	−0.092 (0.288)
Zymotic disease death rate	0.280[b] (0.071)	0.220[b] (0.079)	0.212[b] (0.079)	0.152 (0.092)	0.417[b] (0.065)	0.269[b] (0.070)	0.269 (0.071)	0.206[b] (0.079)
Zymotic × lead	...	...	...	0.169 (0.120)	...	...	...	0.154 (0.100)
Epilepsy death rate	1.300 (1.20)	0.618 (1.12)	1.068 (1.17)	0.829 (1.18)	0.553 (1.42)	0.794 (1.18)	0.825 (1.25)	0.827 (1.16)
Log (population density)	−0.026 (0.037)	0.000 (0.040)	0.004 (0.040)	0.145[b] (0.041)	−0.090 (0.029)	−0.045 (0.028)	−0.044 (0.028)	−0.035 (0.226)
Constant	1.550[b] (0.222)	1.820[b] (0.232)	1.735[b] (0.240)	1.943[b] (0.245)	1.329[b] (0.244)	1.689[b] (0.209)	1.680[b] (0.231)	1.834[b] (0.226)
Adjusted-R^2	0.463	0.484	0.491	0.500	0.589	0.693	0.685	0.704
No. of obs.	42	42	42	42	42	42	42	42

Source: Registrar General (1883).
[a] For precise definitions of the variables, see notes a–i, table B.1.
[b] Significant at the 5 percent level or higher (one-tailed test).
[c] The weighted regression is weighted by births in the county using the aweights algorithm in STATA.

eclampsia, that was driving the increase in deaths from convulsions. Finally, there is some noisy evidence that the effects of lead-solvent water were heightened in those counties where infectious diseases were rampant. This can be seen in regressions (4) and (8), where the lead dummy is interacted with the death rate from zymotic diseases. The interactions are not statistically significant at the 5 percent level, but are significant at the 10–20 percent level, and the coefficients themselves are large.

Appendix C
The Correlates of Lead Solvency

Because the determinants of any given water supply's lead solvency were manifold, the graphs presented in chapter 6 may not adequately capture the true relationships between water lead and the various chemical characteristics of the water supply. But using the data from the Maine State Board of Health (1915), water-lead levels are regressed against measures of hardness, alkalinity, and a series of other chemical constituents thought to influence lead solvency. These other constituents include the levels of chlorine, nitrites, nitrates, ammonia, and albuminoid. The regressions also control for whether the supply came from a public source or a private well, and whether the water sample was from commercially-sold ice. The motivation for including a dummy variable indicating private source is that private water supplies (i.e., those that served a single family or home) often employed much more lead in their piping and pumping systems than did public supplies. To the extent that ice samples were derived from colder water supplies and cold water dissolved less lead than warm water, one expects ice samples to have contained low levels of lead, if all else is held constant.

The results are reported in table C.1, and are generally consistent with what the graphs illustrate. Because alkalinity and hardness are highly co-linear, they are entered separately in two independent regressions (see regression models [1] and [2]). Supporting the Edinburgh and Boston doctrines, the log of alkalinity and the log of hardness are significant predictors of water-lead levels, both statistically and substantively; increases in hardness or alkalinity are associated with reductions in water-lead levels. The log of consumed oxygen is also inversely correlated with water lead, supporting certain aspects of the Boston doctrine. The level

Table C.1
The correlates of lead solvency

Variable	Mean (Std. Dev.)	(1)	(2)
Lead[a]	0.015 (0.062)	dependent variable	dependent variable
Log (hardness)[a]	1.055 (0.903)	−0.015[b] (0.002)	—
Log (alkalinity)[a]	0.363 (1.071)	—	−0.011[b] (0.001)
Log (O_2 consumed)[a]	−2.235 (1.308)	−0.007[b] (0.001)	−0.008[b] (0.001)
Chlorine[a]	1.303 (5.264)	0.001 (0.001)	−0.001 (0.001)
Nitrite[a]	0.001 (0.014)	−0.082 (0.091)	−0.067 (0.091)
Nitrate[a]	0.184 (0.544)	0.017[b] (0.003)	0.014 (0.003)
Ammonia[a]	0.009 (0.149)	0.013 (0.008)	0.014 (0.008)
Albuminoid[a]	0.009 (0.018)	−0.084 (0.081)	−0.086 (0.081)
= 1 if ice	0.017	−0.058[b] (0.010)	−0.048[b] (0.010)
= 1 if private supply	0.791	0.018[b] (0.003)	0.017[b] (0.003)
Constant	—	−0.001 (0.004)	−0.011 (0.004)
Adjusted-R^2	—	0.061	0.059
No. of obs.	2393	2393	2392

Source: Maine State Board of Health (1915).
[a] Measured as parts per 100,000 in water.
[b] Significant at the 0.001 level or higher (two-tailed test).

of nitrates enters both regressions significantly, but the results for nitrates are not robust. If two outlying observations are dropped, this variable is no longer significant; the results for all other explanatory variables remain the same. As expected, the dummy variable for private water supply enters with a positive and significant coefficient, while the dummy variable for ice enters with a negative and significant coefficient. It is also notable that both models explain only 6 percent of the variation in water-lead levels, highlighting the complex and stochastic nature of the underlying relationships.

Notes

Prologue: Exhuming Michael Galler

1. *New York Times*, November 22, 1869, p. 5, and November 28, 1869, p. 8.
2. *New York Times*, November 22, 1869, p. 5, and November 28, 1869, p. 8.
3. *New York Times*, November 20, 1869, p. 2.
4. See *New York Times*, November 30, 1869, p. 2, and December 2, 1869, p. 5.
5. *New York Times*, November 23, 1869, p. 2; November 28, 1869, p. 8; and December 25, 1869, p. 8.
6. The *British Medical Journal* (October 2, 1896, p. 376) reported the following: "A woman named Colvill, residing at Castletown, Isle of Man, has been committed for trial on the charge of attempting to poison her husband with white lead. Suspicions were first entertained by Dr. Wise, the surgeon whom the man consulted. Colvill had during the early part of the year been repeatedly under care for symptoms not easy of explanation, and at length lead was suspected. Dr. Wise very properly communicated with the authorities, and circumstances came out which tended to confirm his opinion. The woman had purchased lead, and had tried to buy arsenic also. It seems probable that, by Dr. Wise's prompt action, a murder has been averted."
7. *New York Times*, December 25, 1869, p. 8.
8. *New York Times*, November 24, 1869, p. 2; December 2, 1869, p. 5; December 22, 1869, p. 2; and December 25, 1869, p. 8.
9. *New York Herald*, December 25, 1869, p. 6. Of course, it is also possible that Galler may have been exposed to the lead from any number of other sources, including beer, wine, food, or perhaps even his occupation. This text, however, focuses exclusively on the possible role of the public water supply. This focus is adopted to help frame the issues raised later in the book, not to exclude other possible sources of exposure.
10. However, after the New York Metropolitan Board of Health issued a short statement encouraging families to flush their pipes thoroughly before drinking the city's water, the *Times* (April 12, 1870, p. 4) did offer the following editorial

statement: "The warning addressed to householders by the Board of Health, in reference to the [city] water, ought not to escape attention. There can be no doubt that the use of lead pipes in houses gives rise to much sickness, for which people are often at a loss to account. The proper plan is to allow the water 'to run off for a few minutes before taking it for drinking or cooking purposes.' This is a very simple precaution and it ought to be dinned into the ears of servants until they take it."

11. Quotations are from *New York Herald*, December 25, 1869, p. 4.

12. For example, as late as the 1890s, physicians were discussing and recommending various antipyretic and antiseptic treatments for typhoid fever. These treatments included the use of chemicals such as sulphurous acid, carbolic acid, iodine, and chlorine. See Stewart (1893), pp. 49–71.

13. For an account of Speer's death, see *New York Times*, January 19, 1890, p. 5.

14. New York City (1869), pp. 420–421.

15. See *New York Times*, August 17, 1936, p. 21, and Quam and Klein (1936).

16. See *New York Times*, September 4, 1992, pp. B1 and B6.

17. For more information on the frequency of elevated lead levels in New York City water in 1992, see *New York Times*, September 4, 1992, pp. B1 and B6. For why lead corrosion subsided over time, see chapter 6.

18. Melosi (2000), p. 83.

19. See Melosi (2000), p. 86.

20. *New York Times*, August 25, 1853, p. 4.

21. See the letter of Dr. Alex E. Hosack dated November 16, 1847, and reprinted in Kirkwood (1859), p. 63.

22. See Kingsbury (1851).

23. See *New York Times*, December 10, 1853, p. 2, and February 4, 1854, p. 1.

24. All quotations in this paragraph are from *New York Medical Gazette and Journal of Health*, September 1853, pp. 408–409.

25. *New York Times*, August 30, 1861, p. 5.

1   The Significance of the Small

1. There is the occasional case of West Nile virus or encephalitis in the United States, but mosquito-related diseases are generally not a major cause of death in wealthy and temperate parts of the world. See Landes (1998), pp. 10–11.

2. On the fragility of certain economies and their associated vulnerability to famines, see Sen (1981).

3. The data on the frequency of lead piping in smaller places are derived from Troesken and Beeson (2003), pp. 187–188.

4. See Chalmers (1940) for the case of a farm family who became gravely ill after drinking water that had passed through a lead pipe three-quarters of a mile long. Lead levels in water drawn from this pipe ranged from 6 to 8 parts per million, 400–533 times the current EPA standard. This case appears to have occurred in Scotland.

5. On the use of lead pipes internationally, see Saunders (1882) and Ingleson (1934), pp. 59–76.

6. See, generally, Grobler, Theunissen, and Maresky (1996). The lead concentrations in the primary teeth of Cape Town residents had a mean level of 109, while secondary teeth had a mean level of 315 ($\mu g/g$). Compare these levels, for example, to those reported in Shapiro et al. (1973).

7. Grobler, Theunissen, and Maresky (1996).

8. The modern EPA standard is 0.015 ppm. Water with a lead level 500 times that standard would contain 7.5 parts of lead per million parts of water.

9. On lead poisoning in the ancient world, see, for example, Swann (1892), Nriagu (1983), and Waldron (1973).

10. See, for example, Needleman (1997, 1998) and Warren (2000), pp. 124–128.

11. See, for example, Preston and Haines (1991), pp. 50–51, who describe the Massachusetts data used here as "quite good." Meeker (1972) concurs with this assessment. See also Vinovskis (1972, 1981). On the quality of the British data, see Wrigley and Schofield (1989), p. 631, who describe England's registration data as "tolerably complete" after 1874, when stiff penalties were introduced for failing to report births and deaths.

12. Lead poisoning more generally construed—that is, including all sources of exposure—still appears to be a serious and global public health problem. See Tong, von Schirnding, and Prapamontol (2000).

13. Anawar et al. (2002) and Paul (2004).

14. Brown and Ross (2002).

15. See, for example, Vincent et al. (2000), Flaten (2001), Durant et al. (1995), Peplow and Edmonds (2004), Puklová et al. (2005), Fowler et al. (2004), Cambell et al. (2004), Hudak (2004), and Liang et al. (2003). Also of interest is a recent study by Gillettte-Guyonnet et al. (2005). These researchers find no evidence that water with a high aluminum level impairs cognitive function among women, but do find evidence that water with high silica content appears to offer protection against cognitive degradation among older women. Although these findings require much more research, one possibility is that silica reduces the lead solvency of water and thereby minimizes lead exposure and its associated neurotoxic effects.

16. Paul (2004).

17. Yoshida, Yamauchi, and Sun (2004). See also "The Tainted Source: Bangladesh's Poisoned Water," *The Economist*, March 29, 2003, p. 54.

18. See, for example, McKay and Moeller (2002) and Davies and Mazumber (2003).

19. See, generally, Haines (2001).

20. Studies emphasizing the increasing importance of the germ theory of disease over the course of the early twentieth century include Ewbank and Preston (1990), Mokyr (2000), Preston and Haines (1991), and Bengtsson et al. (2004).

21. Studies emphasizing the importance of investments in public infrastructure related to health include Preston and van de Walle (1978), McKeown (1976), Cutler and Miller (2004), Cain and Rotella (2001), Meeker (1972, 1976), Brown (1989a, 1989b), van Poppel and van der Heijden (1997), and Szreter (1988, 1997).

22. Murray (1997) and Steckel (1995) provide excellent surveys of the literature and evidence linking changes in nutrition and mortality. See also Haines, Craig, and Weiss (2003) and McKeown (1976).

23. Scientific studies on the long-term evolution of lead exposure include Nriagu (1998) and Budd et al. (2004). Social and medical histories of lead include Warren (2000) and Wedeen (1984). See also Hernberg (2000).

24. See, for example, Troesken (2006a) and Troesken and Beeson (2003).

2   A House for Erasmus

1. Fenner (1850), p. 247.

2. See Fenner (1850), especially pp. 248–254.

3. Recent research suggests that when New Orleans first installed lead water pipes, tap water probably contained excess levels of lead. See Boyd et al. (2004).

4. Fenner (1850). The quotation in the text appears on p. 263 of the article.

5. Fenner (1850), pp. 270–273.

6. Fenner (1850), p. 261.

7. See Baker (1897), p. 274.

8. See Fenner (1850), pp. 263–264.

9. Fenner (1850), pp. 261 and 264.

10. Citing data and quoting passages from a famous French physician, Fenner (1850, p. 259) wrote: "In regard to seasons and climate, Tanquerel shows that of 1217 cases submitted to his observation, the greatest number occurred in June, July, and August, the warmest months of the year. He was convinced that, other circumstances being equal, the number of sick laborers in leadworks was greater during the warm than the cold seasons of the year. He thinks that the heat predisposes to attacks of lead colic, either by favoring the dissemination of lead, or by rendering more permeable the different organs by which it enters the system. He says—'there are some persons who are attacked with lead colic every year, at nearly the same time, although they are employed in the same work from

the beginning of the year to the time when they are taken sick. Summer is generally the season for these periodical attacks.'"

11. Fenner (1850, p. 258) wrote: "The basis of all skepticism that has ever existed in relation to the dependence of this and similar colics, upon lead poisoning, rests entirely on the unsuccessful efforts that were made to detect the lead in a satisfactory manner."

12. See chapters 3, 7, and 8 for discussions of the evolving standard of what was considered a safe water-lead level.

13. See Fenner (1850), particularly p. 267. On the historical understanding of the chemistry of water and lead, see chapter 6.

14. Poincaré (1952).

15. The factors that might predispose some individuals more than others to lead poisoning are discussed in detail later in the chapter.

16. See, for example, Wynne (1911), Schaut (1942), Clement, Seux, and Rabarot (2000), and van Der Leer et al. (2002).

17. See, in particular, Schaut (1942), p. 245.

18. See, for example, United States Environmental Protection Agency (1995); Yiin, Rhoads, and Lioy (2000); and Haley and Talbot (2004), pp. 166–167. On the other hand, Oliveira et al. (2002), show that some of the seasonal variation in blood levels is the result of mobilization of bone-lead stores.

19. See, generally, Needleman (2004).

20. See Lanphear et al. (2000) and Canfield et al. (2003).

21. For a review of the relevant evidence surrounding the search for "threshold" effects with regard to lead's neurotoxicity, see Bellinger (2004). This review essay highlights the importance of synthesizing human and animal studies in ascertaining the "behavioral signature" of lead toxicity.

22. See, for example, Solliway et al. (1994). On the relative immunity of adult neurological systems to lead exposure, see Kuhlman, McGlothan, and Guilarte (1997).

23. This paragraph is based on the following sources: McDonald and Potter (1996), Schwartz (1991), and Sauvant and Pepin (2002).

24. Hamilton (1919), p. 8.

25. Oliver (1897), pp. 962–963.

26. See Oliver (1897), pp. 962–963 and 968–969, and Gowers (1888), pp. 1254–1255. Also, in their textbook *Lead Poisoning and Lead Absorption*, Legge and Goadby dedicate an entire chapter to documenting and explaining the idiosyncratic effects of lead exposure. See Legge and Goadby (1912), pp. 27–43.

27. Onalaja and Claudio (2000).

28. DeMichelle (1984).

29. Hsu and Guo (2002).

30. Bryce-Smith et al. (1977).

31. In a review of the literature, Winder (1993) argues that there is not sufficient evidence to justify the claim that lead exposure can cause birth defects. A more recent review, however, cites much evidence from both animal and human experiments that lead might damage the human chromosome. See Johnson (1998).

32. See, for example, Wibberly et al. (1977). For a review of other studies presenting similar evidence, see Tourmaa (1995). For more recent studies, see Borja-Aburto et al. (1999) and Hertz-Picciotto (2000).

33. See, for example, Bellinger et al. (1991).

34. The idea that fetal and infant insults (exposure to organic and inorganic pathogens) have persistent health effects is called the "Barker hypothesis" or the "fetal-origins hypothesis." See Barker (1994) for a summary and presentation of the hypothesis and some supporting evidence. The Barker hypothesis has been subject to much debate and criticism. For a review of this literature, see, for example, Henriksen and Clausen (2002). Nevertheless, the notion that fetal exposure might be one of the mechanisms that gives rise to a correlation between adult disease and early-life health conditions is intriguing. One of the first researchers to suggest such a mechanism was Tourmaa (1995).

35. See, for example, Hu (1991).

36. See Winder (1993), Ernhart (1992), Ernhart et al. (1986), and Hertz-Picciotto (2000).

37. See Bellinger et al. (1991), Winder (1993), and Ernhart (1992). See also Ernhart et al. (1986), who suggest that low-level intrauterine lead exposure might not even be sufficient to affect the neurological development of the fetus and neonate.

38. See Hertz-Picciotto (2000).

39. Hertz-Picciotto (2000).

40. See Schlemback (2003) for a review and argument that preeclampsia is better thought of as a syndrome with multiple possible causes.

41. See Dawson et al. (2000) and Dawson et al. (1999).

42. Gulson et al. (1997).

43. One might ask why the body mobilizes bone lead along with essential metals like calcium, zinc, potassium, and magnesium. Why, in other words, does the human body not adequately distinguish toxic metals from essential metals? As explained later, lead shares certain biochemical characteristics with both calcium and zinc, and these shared characteristics probably explain the human body's failure to distinguish lead from these other, more desirable metals. The information in this paragraph is based on a study by Dawson et al. (1999). See also Dawson et al. (2000).

Recent animal studies show that low-grade lead poisoning is associated with reduced calcium transport, which in turn generates a physical syndrome in rats that closely resembles preeclampsia. In particular, these studies show that lead alters the growth of aortic smooth muscle cells and that high blood pressure may be related to the metabolic levels of lead and calcium. There is also evidence, from studies of both children and animals, that iron and protein deficiencies might predispose individuals to the toxic effects of lead. See Dawson et al. (1999) for a review of the animal studies. It is also possible that there are interaction effects so that more lead is mobilized in women who are calcium-deficient than in women who are better nourished. The magnitude and functioning of such interaction effects, however, are not entirely clear and require further study. See also Sowers et al. (2002), who present direct evidence of the interaction effects between lead and calcium in an experiment involving 705 pregnant women.

44. See Sowers et al. (2002). Still more evidence on the connection between maternal bone lead and preeclampsia can be found in Rothenberg et al. (2002) and Hu and Hernandez-Avilla (2002).

45. Godwin (2001), p. 206.

46. See the discussion that follows immediately for the relevant citations and supporting evidence.

47. See Knowles, Donaldson, and Andrews (1998) and Todorovic and Vujanovic (2002).

48. The discussion of calcium in this paragraph is taken from Emsley (2001), pp. 84–85.

49. Specifically, lead and calcium are divalent cations (Pb^{2+} and Ca^{2+}). See Onalaja and Claudio (2000) for a more complete discussion of lead's affinity for calcium-dependent enzymes and the significance of the shared ionic structure of calcium and lead.

50. See Godwin (2001), Lidsky and Schneider (2003), Suszkiw (2004), Bressler et al. (1999), and Needleman (2004).

51. See Godwin (2001); Needleman (2004); Suszkiw (2004); Simmons (1993); Markovac and Goldstein (1988); and Thain and Hickman (2001), pp. 209 and 230.

52. Needleman et al. (2002). For those social scientists concerned about unobserved heterogeneity in human testing, animal experiments reveal the same evidence, that is, that lead exposure impairs social functioning and development. See, for example, Bushnell and Bowman (1979) and Rice (1992).

53. For an analysis of the relationship between lead and PKC, see Godwin (2000). On the general functioning of PKC, see Thain and Hickman (2001), pp. 502–503, and Newton (1995).

54. See Needleman and Gatsonis (1990) for a meta-analysis of the relevant modern studies, and Needleman et al. (1990) for a follow-up to Needleman's earlier studies of lead-exposed children and IQ. See also Pocock, Smith, and Baghurst (1994), who review the evidence with a greater focus on the European literature.

55. Specifically, zinc is a divalent cation (Zn^{2+}), as are lead and calcium.

56. Emsley (2001), pp. 495–496, and Godwin (2001).

262   *Notes to Pages 43–47*

57. Emsley (2001), pp. 228–229; Godwin (2001); Onalaja and Claudio (2000); and Needleman (2004).

58. Godwin (2001).

59. Brown et al. (1983).

60. Kim et al. (2002).

61. Barciszewska et al. (2003).

62. Farkas et al. (1987).

63. For a review of the relevant evidence on lead's ability to penetrate the placental barrier, see Tourmaa (1995). One of the few studies to report evidence that the placental barrier is impermeable to lead is Black et al. (2002). This study is based on the comparison of lead levels in hair samples of newborn children and their mothers. It is not at all clear, however, that one should view the amount of lead in hair as the most relevant biochemical marker of undue lead exposure. See also Klein et al. (1994).

64. See, for example, Goyer (1990).

65. See, for example, the discussion in Lidsky and Schneider (2003).

66. On lead's ability to penetrate the BBB, see Bradbury and Deane (1993), Krigman (1978), Moorhouse et al. (1988), and Needleman (2004).

67. Quotations are from Dietert et al. (2004). For additional evidence of lead's effects as an immunotoxicant, see Kuo, Hsiao, and Lai (2001) and Sata et al. (1998).

68. See Johnson (1998) and Landrigan, Boffetta, and Apostoli (2000).

69. See, for example, Oliver (1897).

70. Satarug et al. (2004).

71. See, for example, Nehru and Kaushal (1993) and Jarrar and Mahmoud (2000).

72. Nation et al. (1986), cited in Needleman (2004).

73. See, for example, Papaioannou et al. (1998).

74. There is a dearth of recent research on lead and kidney failure. Nevertheless, a brief review of the existing literature can be found in Brewster and Perazella (2004).

75. Free radicals are molecules with one unpaired electron. Because of this, these molecules draw electrons from other molecules and thereby cause oxidation (rusting). See Sharp (2003), p. 177, for a formal definition and description.

76. Oktem et al. (2004).

77. Aykin-Burns et al. (2003).

78. Demasi et al. (1996).

79. Gurer-Orhan, Sabir, and Ozgünes (2004).

80. See, for example, Page (2002).

81. See Elwood, St. Leger, and Morton (1976) and Lauwerys et al. (1977). See also Needleman et al. (1984), who fail to find any correlation between water lead and lead in umbilical cord blood.

82. See Moore et al. (1979) and Moore et al. (1982).

83. See, for example, Elwood, St. Leger, and Morton (1976).

84. See, for example, Gloag (1981), Meyer et al. (1998), Schütz et al. (1997), and Kahn et al. (2001).

85. It seems likely that most in utero, lead exposure is related to tap water ingested by the mother. In infancy, recent research suggests lead-contaminated tap water used in making baby formula is important. See Shannon and Graef (1992). As children grow, soil and household dust become more important. See Lanphear et al. (2002). Among non-occupationally exposed adults in urban areas, tap water and leaded gasoline are important sources of exposure. See Sartor and Rondia (1980). Lanphear and Roghmann (1996) report evidence suggesting that time spent outdoors influences lead uptake from soils.

86. The articles discussed in this paragraph are Delves and Campbell (1993), Bois et al. (1989), and Lanphear et al. (2002). See also Thomson et al. (1989).

87. On the importance of complex, nonlinear relationships in health outcomes, see, generally, Schwartz (1993). For lead in particular, see, for example, the preceding discussions of lead and fetal development, and the work of Bruce Lanphear and others, which shows that the marginal impact of lead might be highest at the lowest levels of exposure. See also the discussion above regarding the work of Michael Moore, who identified a nonlinear relationship between water lead exposure and blood lead.

3 Fixing Alice

1. Alice is case I in Pope (1893). On the growing use of lead plaster as means of birth control and abortion during this period, see also Wrangham (1901), Hall (1905), Branson (1899), Hall and Ransom (1906), and Ransom (1900). For a modern essay that places the use of diachylon in its broader historical context, see Sauer (1978).

2. Anne's case is based on the discussion in Pope (1893).

3. Pope (1893), p. 10.

4. See Pope (1893), p. 10.

5. Mary's history is that of case I reported by Wrangham (1901), p. 72. A case reported by Branson (1899) also suggests that one could make a large number of pills from a small amount of diachylon. In particular, Branson's case involved a lead-poisoned woman who had fashioned "some sixty pills" from a bulk amount of diachylon "about that of one's thumb." She took all of the pills over the next two days, and developed a severe case of lead poisoning. While Branson worked in Edgbaston, he noted press reports indicating that the identical use of lead plaster was taking place in Birmingham.

6. Charlotte's history corresponds to case II reported in Ransom (1900), p. 1590.

7. Hall (1905), p. 585. The use of diachylon for ostensibly medical purposes has continued. For example, a comparatively recent report in the German medical literature describes the case of a sixty-four-year-old female who had used a diachylon ointment to treat "extensive bilateral leg ulcers." After using the ointment for more than a year, the patient developed lead poisoning with "general weakness, loss of weight, anemia, hypotension, and neuropathy." See Bilonczyk, Partsch, and Donner (1989).

8. See Massachusetts State Board of Health (1900). The following section builds heavily on Troesken (2006b).

9. Letting L equal the water-lead concentration expressed in terms of parts per 100,000, the abortifacient equivalent can be expressed as

$AE = 0.878/L.$

The derivation for this formula is provided in the notes to table 3.2.

10. See, generally, Ingleson (1934).

11. Quam and Klein (1936).

12. See, for example, *Milton News*, February 15, 1902, p. 1. The paper reported that the chairman of the State Board of Health, a Dr. Walcott, stated: "The use of lead pipe for conveying water ought to be discouraged generally, as there is always danger of poisoning from it. *The newer the pipe, the greater the danger*" (emphasis added). See also Massachusetts State Board of Health (1895), pp. xxiv and 30–31. For a discussion of other potential determinants of water lead levels besides age of pipe, water hardness, and CO_2 levels, see chapter 6. The inordinately high lead levels observed in Massachusetts drinking water did not arise solely from the use of lead service pipes. There were plenty of other sources of lead in tap water, including the pollution of water sources, and lead-based pipes and solders used for the interior plumbing of homes. Lead service pipes, however, were the primary source of lead in drinking water. See Quam and Klein (1936), p. 779, particularly Tables I and II, and Massachusetts State Board of Health (1900), pp. 491–497.

13. Swann (1889).

14. Swann (1889).

15. Swann (1892), p. 194.

16. For all quotations in this paragraph, see Swann (1892), p. 194. Ironically, there is evidence to suggest that lead poisoning might have been common in the ancient world. Nriagu (1983, 1998) argues that lead and lead poisoning were pervasive in the ancient world, while Waldron (1973) and others have claimed there is insufficient evidence to justify such a position.

17. The one exception to this appears to be the literature on leaden abortion pills presented at the beginning of the chapter. But even here it is clear that it was not until 1905, with publication of Hall's article, that physicians became aware of how little lead was needed to induce abortion and disrupt menstruation.

18. Stainthorpe (1914).

19. See Porritt (1931) and Milligan (1931).

20. See appendix A for a detailed discussion of how to address these issues.

21. In a short essay on childhood lead poisoning, Holt (1923, p. 232) wrote: "Only a few reports could be found in the literature in which the placental transmission of lead was beyond question. Ganyaire demonstrated the presence of lead in the placenta, liver and brain of an infant dying when 48 hours old. Legrand and Winter report a similar case." But such cases, according to Holt, were few and far between.

22. Chapter 2 reviews the evidence regarding lead's many toxic mechanisms.

23. See Bell, Hendry, and Annett (1925) and Mitsui (1934).

24. These studies are cited and explained in Dilling and Healey (1926).

25. For the quotations in the paragraph, see Oliver (1911), pp. 84–85. Another expert who characterized lead as a "race poison" was Alice Hamilton, a physician initially affiliated with the U.S. Government and then later with Harvard University. See Hamilton (1925), pp. 110–111, and particularly the title of her chapter 8.

26. Oliver (1911), pp. 83–84.

27. Oliver (1911), pp. 83–84.

28. Oliver (1911), pp. 84–85.

29. Oliver (1911), pp. 87–88.

30. Oliver (1911), pp. 83 and 93.

31. Schettler et al. (2000).

32. Boston Water Commissioners (1848), pp. 49–52.

33. Massachusetts State Board of Health (1899), pp. 490–491.

34. Greene (1889), p. 533.

35. See Greene (1889), p. 534, and Hills (1894).

36. See Putnam (1887, 1889). See also the discussion of Putnam's work in chapter 5.

37. Thresh and Beale (1925), p. 162.

38. See Thresh and Beale (1925), p. 162. See also Thresh (1905). This article documented a case of lead poisoning in an area with hard water supplies. The victim's water had a lead content of 0.3 and 0.65 grains per gallon from two samples apparently taken after normal use. Thresh himself measured samples from water that had stood in the lead pipes and pump overnight. These two samples contained 1.8 and 1.4 grains of lead per gallon of water, 2,052 and 1,596 times greater than the modern EPA standard.

39. This paragraph is based on Baker (1897), p. 51; Weston (1920); *The Sanitarian*, July–December, 1899, pp. 230–232; and Massachusetts State Board of Health (1900), pp. xxxii–xxxiii.

40. See chapter 6 for a discussion of how water treatment processes can affect the lead solvency of water supplies.

41. On changes in the lead levels in the Milford-Hopedale water supply, see Weston (1920). On the standards adopted by the Massachusetts State Board of Health as to a safe amount of lead in the water, see Committee on Service Pipes (1917), pp. 355–357.

42. Weston (1920).

43. Weston (1920).

44. See the prologue and chapter 8 for discussions of New York City and Glasgow.

4   The Latent History of Eclampsia

1. Waters (1894), p. 682.

2. Waters (1894), p. 682.

3. Waters (1894), p. 682.

4. See Porritt (1934), pp. 3–4, and Kerr (1933), pp. 201–203. The quotation is from Young, Sym, and Crowe (1932), p. 1237, as are some of the data on subsequent disability.

5. See Grulee (1907) and Porritt (1934).

6. On the importance of abdominal pressure, see, for example, Nash (1892). Nash believed that the developing fetus impaired the mother's ability to excrete urine, which, in turn, resulted in the maternal body becoming overwhelmed by toxins.

7. For a brief review of the theories of eclampsia and associated evidence, see Porritt (1934), pp. 2–9.

8. As quoted by Porritt (1934), p. 36.

9. As quoted by Porritt (1934), p. 36.

10. See, for example, Hiden (1912), Murray (1911), Townsend (1865–1866), and Rea (1894–1895).

11. Murray (1911), p. 187.

12. Hiden (1912). Whatever the parallels between eclampsia and chloroform poisoning, the use of drugs to treat eclampsia was a dangerous proposition in the nineteenth century. See, for example, Townsend (1865–1866), who described the case of an eclamptic patient killed by his efforts to treat her with opium. See also Rea (1894–1895), who described the use of morphine and chloroform and the not-infrequent side effects of such treatments. Similarly, a physician at Johns Hopkins University, writing in 1935, argued that the older, more radical treatments for eclampsia were less effective, and perhaps more dangerous, than newer and milder treatments. See Peckham (1935).

13. See Young (1914).

14. For authors expressing views sympathetic to the renal insufficiency theory, see Rea (1894–1895), Hill (1911), Hale and Cantab (1892), Young and Miller (1920), and Sylvester (1899–1900).

15. See Porritt (1934), pp. 28–31, for a review of the evidence.

16. See Theobald (1930).

17. Theobald (1930), p. 1116.

18. Theobald (1930), p. 1116.

19. The idea that poorly digested food created poisons and ultimately led to eclampsia was not original to Theobald. One of the primary sources of evidence for this "poison food" theory was in the case of Germany during World War I, when in the midst of widespread starvation and food shortages, eclampsia rates fell while the death rates from other diseases rose sharply. See Tweedy (1919).

20. Waters (1894), p. 682.

21. Waters (1894), p. 682. Significant publications on eclampsia and maternal mortality published during the early twentieth century do not mention the Waters article, not even those that expressed sympathy with the idea that there could be connection between lead and eclamplsia. See, for example, Porritt (1934) and Kerr (1933).

22. Porritt (1934), pp. 9–11. See also Troesken (2006a).

23. Porritt (1934), especially pp. 9 and 67–70.

24. Porritt (1934), pp. 69–70, replaced his lead pipes with pipes of block tin, but he was a relatively wealthy man. See chapter 7 for a discussion of the costs and benefits of block tin or tin-lined pipes.

25. See Porritt (1934), pp. 11, 39, and 68.

26. See Porritt (1934), pp. 2, 15–16, and 40–41.

27. The quotations and information in this paragraph are from Porritt (1934), pp. 40–42.

28. Porritt (1934), p. 31.

29. Compare, for example, Porritt (1934), pp. 31–36, with the modern literature on lead poisoning discussed in chapter 2, or the symptoms described in chapter 5. Porritt, it should be emphasized, was stylizing his picture of eclampsia to maximize its similarity with lead poisoning. When other physicians described the symptoms of eclampsia and preeclampsia, they too painted a picture that was similar to that of lead poisoning. Consider the description of eclampsia offered by Kruetzmann (1898), p. 677, and the descriptions of lead poisoning described by Brown in chapter 5. Compare, as well, the cases of peculiar and fatal convulsions in four children born to a lead-poisoned father as reported by Thomson (1923) to the childhood pathologies described in Grulee (1907); all of Grulee's cases involved children born to eclamptic mothers.

30. See Porritt (1934), p. 66. Chapters 3 and 6 and appendices A and C present evidence linking water-lead levels to water softness and acidity.

268 Notes to Pages 87–103

31. Porritt (1934), especially pp. 52–55.

32. Porritt argued that deaths from sepsis outnumbered deaths from eclampsia by a factor of 2, but other data sources suggest a larger number. See Porritt (1934), p. 2.

33. Porritt (1934).

34. Porritt (1934), pp. 5–6.

35. Porritt (1934), p. 58.

36. All quotations in this paragraph come from Porritt (1934), p. 47.

37. Porritt (1934), pp. 57–58.

38. Porritt (1934), pp. 57–58.

39. See Porritt (1934), pp. 2–3 and 77–80.

40. Furthermore, it is clear from this figure that adjusting the functional form of the underlying econometric model will not change this result. For any given rate of sepsis, eclampsia rates are consistently higher in boroughs with lead-solvent water than in boroughs without.

41. Cited as Registrar General (1883).

42. On the ability of other physiological changes to mobilize bone lead, see, for example, Han et al. (1999). Using rats in a controlled environment, this study shows how sudden weight loss and exercise can cause an increase in bone-lead mobilization.

5 The Secret of Dr. Porritt's Society

1. Porritt (1931), p. 92.

2. Porritt (1931), p. 93.

3. Porritt (1931), p. 93.

4. Porritt (1931), p. 93.

5. Porritt (1931), p. 92.

6. See Cullen, Robins, and Eskanzi (1983) and Mesch, Lowenthal, and Coleman (1996).

7. Mesch, Lowenthal, and Coleman (1996).

8. Porritt's books included two novels, *The Factory King* and *Cornered*; a textbook on thoracic surgery; and three medical monographs, one on religion and medicine, one on the dangers of drugs, and the other on the abdomen in pregnancy. Porritt's last book, on eclampsia, was discussed in chapter 4. See Porritt (1883, 1891, 1895, 1916, 1923, 1926, 1934).

9. See Baker (1786), pp. 420–421.

10. Baker (1786), pp. 421–423.

11. Baker (1786), pp. 421–423.

12. Alderson (1839), pp. 87–88.

13. Alderson (1839), pp. 89–90.

14. Lindsay (1859), pp. 23–24.

15. Thomson (1882), pp. 116–117.

16. Rees and Elder (1928), p. 715, citing Hamilton (1925), pp. 55–57.

17. This brief review is taken from Ingleson (1934), p. 4. See also Warren (2000), pp. 24 and 35–37. See, however, Burnham (2005) who argues that undiagnosed lead poisoning was not widespread.

18. Centers for Disease Control and Prevention (2001).

19. See Centers for Disease Control and Prevention (2001).

20. *New York Times*, July 5, 1977, p. 5.

21. *New York Times*, July 5, 1977, p. 5. Other examples of doctors in the late twentieth century mistakenly attributing lead-related pathologies to other causes are discussed in Beigel, Ostfeld, and Schoenfeld (1998); Perleman et al. (1993); and Smitherman and Harber (1991).

22. Harris (1918), pp. 140–141. For additional statements on the problems with using urine-lead levels to diagnose lead poisoning, see Lowndes (1936) and Smith, Rathmell, and Marcil (1938).

23. On the symptomology of lead poisoning, and the use of symptoms in diagnosing lead poisoning, see Smith, Rathmell, and Marcil (1938), especially pp. 472–474.

24. See Stainthorpe (1914), p. 213.

25. See Dana (1848), pp. 12 and 53, and Aub et al. (1926), p. 23.

26. See Smith, Rathmell, and Marcil (1938), p. 473.

27. Dana (1848), p. 16.

28. Aub et al. (1926), p. 162.

29. See Linenthal (1914), p. 1796.

30. Harris (1918), pp. 140–141.

31. See Stainthorpe (1914), p. 213.

32. Holt (1923).

33. See, for example, Dana (1848), pp. 12–15, and Stainthorpe (1914).

34. Linenthal (1914), p. 1796.

35. See Bell, Hendry, and Annett (1925). See also the related discussion in chapter 2.

36. This case is described by Wright, Sappington, and Rantoul (1928), p. 237.

37. Linenthal (1914), p. 1297.

38. See, for example, the reaction to Dr. Kingsbury's paper discussed in the prologue. See also *New York Times*, December 10, 1853, p. 2.

39. Bramwell (1931), p. 92. Bramwell also documented cases of lead encyphelopathy being mistaken for a brain tumor, and lead poisoning simulating gastric ulcer and progressive muscular atrophy.

40. See Myers, Gustafson, and Throne (1935), pp. 579–580.

41. On the disease profile of nineteenth-century cities, see, for example, Preston and Haines (1991), Ferrie and Troesken (2005), Meeker (1972, 1976), Troesken (2004), Costa (2000), Bengtsson et al. (2004), and Haines (2001).

42. This paragraph is based on Evans (1988), p. 127; McNeill (1977), pp. 266–267; and, more generally, Rosenberg (1971).

43. Troesken (2004), pp. 65–66.

44. See Putnam (1887, 1889).

45. See Putnam (1889), p. 530. See also Putnam (1887).

46. For example, reviewing studies of lead exposure among the general population conducted during the 1920s and early 1930s, Lowndes (1936, p. 44) wrote: "One important fact has been established by these investigations, namely, that the occurrence of lead in the body is the rule rather than the exception." While opinions differed as to "whether this small amount of lead should be regarded as normal or an accidental constituent of the body," Lowndes claimed that "for all practical purposes it is usually referred to as 'normal.'" For a more general discussion of the view that "some lead was normal," see Needleman (1998) and Warren (2000), pp. 7, 204, and 211. Needleman's essay describes the intellectual battle between Robert Kehoe (a scientist affiliated with the lead industry who promoted the normalization view) and Clair Patterson (who promoted the opposite). As Needleman's own work has shown, some lead is not normal, and the first unit of exposure might well induce the most incremental harm. See also Lanphear et al. (2000). This issue is discussed and documented further in chapter 2.

47. See Putnam (1889), pp. 532–533.

48. Wright, Sappington, and Rantoul (1928).

49. Wright, Sappington, and Rantoul (1928), pp. 234–235.

50. Wright, Sappington, and Rantoul (1928), pp. 239–242.

51. Wright, Sappington, and Rantoul (1928), pp. 239–242.

52. Wright, Sappington, and Rantoul (1928).

53. Wright, Sappington, and Rantoul (1928).

54. See Brown (1889), pp. 26–27, and a short summary of Dr. Brown's work found in *British Medical Journal*, January 18, 1890, p. 138.

55. Brown (1889).

56. See Brown (1889), especially pp. 18–25.

57. See, for example, Brown (1889), p. 10.

58. Brown (1889), p. 5.

59. See, generally, Needleman (2004). Chapter 2 also discusses this issue when considering the mechanisms that underlie lead's neurotoxic effects.

60. See Hackley and Katz-Jacobson (2003) and Tourmaa (1995) for reviews of the evidence that lead can be transmitted through breast milk and influence the health outcomes of nursing infants.

61. See Ferrie and Troesken (2004) for a discussion of this issue and a brief review of relevant studies.

62. Brown (1889), p. 10.

63. See chapter 2 for a discussion of the factors that influence individual vulnerability to lead poisoning and the associated evidence.

64. For this fascinating rat experiment, see Schneider et al. (2001). There is an important caveat to applying this study to Dr. Porritt and other adults: The study focused on young, developing rats, and the effects of lead as a neurotoxin are much greater on the young than the old.

6   A False Sense of Simplicity

1. See Adams (1852), p. 171. Adams later wrote: "Yet many well-informed physicians, even at the present day, are not roused by [lead's] dangers, and probably will not be, until the frequency of disease caused by it, attended often by fatal termination, but still oftener by loss of muscular power, which renders life almost a burden, has spoken in tones too loud to be passed unheeded, and under circumstances too afflictive to be longer resisted."

2. See Adams (1852), pp. 168–169.

3. Part of Christison's original 1845 essay is reprinted in Kirkwood (1859), pp. 302–319.

4. On these issues, see Lyhus (1989), Hopwood et al. (2002), Garrett (1891), Whipple (1913), and Weston (1920).

5. See Boston Water Commissioners (1848) and Adams (1852), pp. 167–169.

6. Adams (1852), pp. 165–168.

7. Adams (1852), especially p. 166.

8. Nichols (1860), pp. 149–150.

9. See Cardew (2003), p. 2830, Figure 6.

10. See, for example, the many experiments conducted by Garrett (1891). See also Bunker (1921), who explored the lead solvency of a single water source over time and showed that it could vary greatly.

11. See Halem et al. (2001), Cardew (2003), and Gregory (1993).

12. See, for example, Ingleson (1934), pp. 27–58 and 76–92; Cox (1964), pp. 174–210; Davidson et al. (2004); van Der Leer et al. (2002); and Clement, Seux, and Rabarot (2000).

13. On the origins and incidence of lead poisoning in areas with soft moorland water supplies, see Local Government Board (1888–1889), pp. 339–357 and 453–476; Kirker (1890); and Ingleson (1934), pp. 55–68. On the incidence of

water-related lead poisoning in the north of England at this time, see the unsigned editorial comment ("Lead Poisoning") in *British Medical Journal*, November 8, 1890, p. 1089. On the role of rainfall, or the lack thereof, see Brown (1889, p. 3), where he wrote: "Very few towns have escaped where the public water supply is largely dependant [sic] upon the rainfall."

14. See, for example, Lyhus (1989), Maugh (1984), and Wilson (1979).

15. See Brown (1889), pp. 26–27; Tattersall (1897); Kirker (1890), p. 71; and Allen (1882), p. 145. Brown (1890, p. 420) wrote: "Another source...of the increased acidity in water may be the vast quantities of free sulphuric acid in the air due to the increased combustion of coals. The [sulphuric acid] is washed down by the rain, which is then collected from the moors."

16. White (1889), p. 459. Writing in 1922, Thresh explained why it was so difficult to identify the source of lead solvency in moorland water. Specifying two acids—quinic acid and humic acid—he wrote that "these acids are worthy of especial mention, as without their aid the action of most moorland waters on lead could not be imitated in artificially prepared waters." If scientists in the nineteenth century had not happened to strike on these acids in their laboratory experiments, they would have floundered in their experiments. See Thresh (1922), p. 466.

17. See, generally, Gregory (1993) for a straightforward and relatively accessible analysis of the factors determining lead solvency, including water hardness, alkalinity, and pH.

18. Nichols (1860), p. 149.

19. Thresh (1905), pp. 1033–1034. Thresh specified water from the Bageshot Sands as a hard water supply that was also lead solvent. On the complex chemistry and geology that gave rise to lead solvency in waters from the Bageshot, see Irving (1883, 1885).

20. See Lindsay (1859) and Wilson (1966).

21. Rees and Elder (1928).

22. Regressing hardness against alkalinity confirms visual inspection:

$Hardness = 0.880 + 1.397 * Alkalinity$
t-statistic   (12.6)   (87.6)
$R^2 = 0.714, N = 3079.$

23. Two caveats are in order, though. First, hindsight is always 20/20. Second, the fact that alkalinity is a near-perfect predictor of lead solvency in this sample does not imply that samples taken from other regions with different geophysical characteristics might not generate different patterns.

24. The Massachusetts data are described in chapter 3, and are used because Maine did not report information regarding the level of carbonic acid in water supplies.

25. The emphasis is added. See Billings (1898), p. 68.

26. See the discussion in chapters 3 and 7. See also the various reports of Massachusetts State Board of Health (1895, 1896, 1898, 1899, 1900, 1905).

27. See Tattersall (1897), Inglis (1910), Thresh and Beale (1925), Smith and Chaplin (1904), Smith (1910), Frankland (1890), and Reade (1921).

28. This remains true today, when there are multiple techniques of limiting water-lead exposure, and choosing among these techniques is contingent on "site-specific conditions." See Boyd et al. (2000).

29. On Sheffield's search for the optimal treatment process, see Ingleson (1934), pp. 33–34. For the buildup of a bacterial slime with the use of chalk, see Whipple (1913).

30. See Local Government Board (1913–1914), particularly Appendix A, Number 7, titled "Dr. Frank Seymour's Report on the Occurrence of Lead Poisoning in the Urban District of Guisborough, and Its Relation to the Public Water Supply." For a brief synopsis of the events in Guisborough, see Ingleson (1934), pp. 64–65.

31. On the experience of Huddersfield, see Ingleson (1934), pp. 62–63.

32. For the history of Wakefield's water treatment efforts, see Smith (1910); Smith and Chaplin (1904); and Ingleson (1934), pp. 65–66. For numerous other examples of the ongoing efforts to monitor and minimize water plumbism, see Local Government Board (1908–1909), especially Appendix A, Number 10, titled "Water Supplies Shewing Capacity to Act on Lead."

33. On the per-unit and capital costs of water treatments designed to prevent lead poisoning, see, for example, Frankland (1890), p. 240; Inglis (1910); and Smith (1910). Often water suppliers did not even need to purchase any capital equipment to begin treating their water. Workers simply added a little lime and/ or chalk to the water reservoirs prior to distribution through street mains.

34. Writing in 1905, Allen Hazen characterized the costs of water filtration systems this way: "Occasionally, with a very easily [filtered] water, and with conditions favorable for cheap construction, the cost may be as low as $6 or $8 per million gallons. On the other hand, with waters which are difficult to [filter], or where the conditions of construction are difficult, the cost may be increased to $15 or even $20 per million gallons. In a general way, the purification of the water adds from 10 to 20% to the entire cost of furnishing and supplying water to an American city." See Hazen (1905), p. 149.

35. I base this figure on an informal survey of acquaintances in the plumbing and building trades.

36. These towns included Andover, Brookfield, Chicopee, Cohasset, Easton, Haverhill, Lawrence, Malden, Melrose, Nantucket, New Bedford, Newton, North Attleborough, North Brookfield, Plymouth, Revere, Sharon, South Hadley, Swampscott, and Wakefield.

7   Responsibility in the Court of the Absurd

1. For the facts of this case, see the article "The Huddersfield Lead Poisoning Case," in *Chemical News*, August 25, 1882, p. 88, and *Milnes v.*

Huddersfield 10 QBD 124 (1882). One analyst found only 0.01 grains of lead per gallon, a finding which was clearly an outlier given the findings of the other chemists. See *Chemical News*, August 25, 1882, p. 88.

2. For quotation, see the comments of Lord Blackburn in *Milnes v. The Borough of Huddersfield*, 11 App. Cas. 511 (1886). See also *Milnes v. the Mayor, &c., of Huddersfield* 12 Q.B.D. 443 (1883) and *Milnes v. the Mayor, &c., of Huddersfield* 10 Q.B.D. 124 (1882).

3. See *Milnes v. Huddersfield*, 11 App. Cas. 511 (1886); Waterworks Clauses Act 1847 (10 & 11 Cit. C. 17) s. 35; Waterworks Clauses Act 1863 (26 & 27 Vict. C. 93) s. 19; and MacMorran and Scholefield (1914), pp. 126–129.

4. See *Milnes v. Huddersfield*, 11 App. Cas. 511 (1886), particularly the comments of Lords Blackburn and Bramwell.

5. See *Milnes v. Huddersfield*, 11 App. Cas. 511 (1886), particularly the comments by the Earl of Selborne and Lord Blackburn.

6. See *Milnes v. Huddersfield*, 11 App. Cas. 511 (1886). The quotations here are from the comments by the Earl of Selborne and Lord Blackburn.

7. This paragraph is based on articles in the following issues of *Milton News*: January 11, 1902, p. 1; February 1, 1902, p. 1; and February 15, 1902, pp. 1 and 4. See also *Welsh v. Milton Water Company*, 200 Mass. 409 (1909).

8. See *Milton News*, January 11, 1902, pp. 1–2; February 1, 1902, p. 1; and February 15, 1902, pp. 1 and 4.

9. *Welsh v. Milton*, 200 Mass. 409 (1909), pp. 409–410.

10. Aside from the odd assignment of liability, the trial judge's ruling was blind to recent history. Between 1898 and 1900, there were several epidemic-like outbreaks of water-related lead poisoning in numerous cities and towns in Massachusetts. These outbreaks were documented by the state board of health, and enjoyed wide discussion among engineers and water company superintendents.

11. *Welsh v. Milton*, 200 Mass. 409 (1909), especially pp. 410–411.

12. After 1930, courts in England and the United States began to rule in favor of consumers, and hold water suppliers liable for cases of water-related lead poisoning. See *Barnes v. Irwin Valley Water Board*, 1 K.B. 21 (1939) *Earle R. Horton v. Town of North Attleboro*, 302 Mass. 137 (1939). There is, however, evidence to suggest that over the past few decades, courts and regulators have begun rehabilitating nineteenth-century law, and forcing consumers to bear an increasing share of the liability for lead contaminated water supplies. For recent legal developments in the United States, see Bellia (1992), and the following cases: *American Water Works Association (AWWA) v. Environmental Protection Agency (EPA)*, 40 F.3d 1266 (1994); *Bass v. Ledbetter*, 257 Ga. 738 (1988); and *Environmental Defense Fund v. Douglas M. Costle*, 578 F.2d 337 (1978). For legal and regulatory developments in the United Kingdom, see Jones (1989).

13. Kirby (1896), p. 125.

14. Dana (1848), p. 420.

15. For more details on the characteristics and problems associated with tin-lined pipe, see *Popular Science Monthly*, June 1888, p. 281. This short essay summarizes work by British chemists and engineers on the difficulties with tin-lined pipe.

16. Committee on Service Pipes (1917), pp. 326–327.

17. See *New York Times*, January 6, 1870, p. 5.

18. Porritt (1934), pp. 69–70. For a thorough review of the engineering literature on the efficacy of tin-lined lead pipes, see Ingleson (1934), pp. 10–13.

19. Kirby (1896), p. 125, and Ingleson (1934), pp. 10–13.

20. Ingleson (1934), pp. 11–12.

21. Porritt (1934), p. 70.

22. Brown (1889), p. 30.

23. See Massachusetts State Board of Health (1900).

24. The quotations in this paragraph are from Gallagher (1885).

25. Gallagher (1885) quoted the following passage: "He showed me, and behold the Lord stood upon a wall made by a plumb-line, with a plumb-line in his hand; and the Lord said unto me: 'Amos what seest thou?' And I said, 'A plumb line.' Then said the Lord: 'Behold, I will set a plumb-line in the midst of my people Israel; I will not again pass them by anymore.'"

26. Gallagher (1885).

27. Gallagher (1885).

28. Gallagher (1885).

29. See *Engineering News*, September 16, 1916, pp. 595–596.

30. See Committee on Service Pipes (1917).

31. *New York Times*, June 30, 1870, p. 5.

32. See *New York Times*, January 6, 1870, p. 5.

33. See *New York Times*, August 25, 1853, pp. 4–6.

34. See *New York Times*, August 29, 1853, p. 5.

35. *New York Medical Gazette and Journal of Health*, September 1853, p. 409.

36. *New York Medical Gazette and Journal of Health*, September 1853, p. 409.

37. See Quam and Klein (1936), and discussion in the prologue.

38. Compare Ellet's discussion to Sedgwick (1901), and the broader review contained in chapter 6.

39. All quotations in this paragraph are from *New York Medical Gazette and Journal of Health*, September 1853, pp. 408–409, and from the letter, "Remarks on Leaded Pipes, With an Account of a Cheap Method of Rendering Leaden Pipe Harmless," in *New York Medical Gazette and Journal of Health*, September 1853, pp. 465–466. The editorial comments following this letter appear to have been written by Reese, who edited the journal.

40. In contrast to the cod liver oil and the atmospheric plates, trusses were in general an effective means of treating hernias. For evidence on the efficacy of trusses, see Song and Nguyen (2003).

41. For the truss advertisement, see *New York Times*, April 15, 1853, p. 8.

42. On Reese's position in the American Institute, see *New York Times*, October 30, 1856, p. 8, and December 7, 1855, p. 4.

43. *New York Times*, September 30, 1853, p. 5.

44. *New York Times*, September 30, 1853, p. 5.

45. *New York Times*, September 30, 1853, p. 5.

46. See the following issues of the *New York Times* for details of the battle between Draper and Reese: April 12, 1855, p. 4, and April 13, 1855, p. 4.

47. *New York Times*, July 16, 1853, p. 3.

48. *New York Times*, May 7, 1858, pp. 1–2.

49. *New York Times*, May 7, 1858, pp. 1–2.

50. The quotation is from *Lowell Daily Sun*, December 16, 1893, p. 4. On the outbreak of water plumbism in England during the 1880s and 1890s, see the following unsigned editorials and articles in *British Medical Journal*: "Lead Poisoning at Sheffield," January 21, 1888, p. 137; "Lead in Urban Water Supplies," June 22, 1889, p. 1414; "Lead in Public Water Supplies," January 18, 1890, pp. 139–140; "Lead in Public Water Supplies," February 22, 1890, p. 439; and "Lead Poisoning from Drinking Water," August 23, 1890, p. 471. See also the following unsigned editorials and articles in *Lancet*: "Lead Poisoning by Potable Water," October 12, 1889, p. 753, and "Lead in a Lancashire Water," July 11, 1908, p. 120. On the incidence of water-related lead poisoning in England at this time, see the unsigned editorial comment, "Lead Poisoning," in *British Medical Journal*, November 8, 1890, p. 1089. The upshot of these many citations is that there was an easily documented history of water plumbism, published in widely read and well-respected outlets.

51. See Massachusetts State Board of Health (1871). See also Massachusetts State Board of Health (1878).

52. Massachusetts State Board of Health (1871), p. 23.

53. Massachusetts State Board of Health (1871), pp. 23–25.

54. Massachusetts State Board of Health (1900), pp. xxxi–xxxix.

55. Massachusetts State Board of Health (1900), pp. xxxi–xxxix.

56. Quotations found in Committee on Service Pipes (1917), p. 356.

57. *Lowell Daily Sun*, January 24, 1894, p. 1, and March 13, 1894, p. 1.

58. See Allen (1888).

59. Allen (1888), pp. 356–357.

60. See, generally, Dana (1848), pp. 363–422. In an appendix to his book, Dana presents several statements on the dangers of using lead pipes in Lowell.

61. Kirkwood (1859), p. 153.

62. Kirkwood (1859), p. 154.

63. Kirkwood (1859), pp. 154–155.

64. See Dana (1848), pp. 363–422, and the reports found in Kirkwood (1859), pp. 66–67, 137–144, 153–167, and 313–315.

65. Massachusetts State Board of Health (1871), p. 24.

66. Massachusetts State Board of Health (1871), p. 31.

67. Massachusetts State Board of Health (1871), p. 27. Other examples of water plumbism can be found in Massachusetts State Board of Health (1895, 1896, 1898).

68. See Whitsell (1941), p. 995.

69. Whitsell (1941), p. 995.

70. Whitsell (1941), p. 995.

71. See, for example, *Milnes v. Huddersfield*.

8   The Legend of Loch Katrine

1. The quotations are from Clarke (1928), p. 5.

2. Burnet (1869), pp. 2–3, and Clarke (1928), p. 7.

3. Burnet (1869), p. 2, and Clarke (1928), p. 7.

4. Cage (1983), p. 179.

5. Burnet (1869), pp. 2–3. For a more recent survey of disease and sanitary conditions in Glasgow during the early nineteenth century, see Lees (1996).

6. Burnet (1869), pp. 5–8, and Clarke (1928), p. 8. On the propensity of private water companies to underserve poor communities, see Troesken (2001).

7. Burnet (1869), pp. 8–10.

8. Burnet (1869), pp. 14–15.

9. Burnet (1869), pp. 32–41.

10. Clarke (1928), p. 10.

11. For the relevant population statistics, see Burnet (1869), pp. 15–17, and Flinn (1977), p. 302.

12. Cowan (1840), p. 261. On the forces driving Irish emigration during the nineteenth century, see Guinnane (1997). On conditions in Ireland before the famine, see O'Gráda (1999), pp. 24–34.

13. Cage (1983).

14. Burnet (1869), p. 15, and Clarke (1928), pp. 8–11.

15. See Flinn (1977), pp. 377–378; Szreter and Mooney (1998), p. 96; and Glass (1964). Some early observers attributed Glasgow's declining health to the influx

of immigrants, who were predominantly poor, underfed, and unhealthy. See, for example, Cowan (1840), pp. 261 and 275. While the relative health status of new immigrants likely explained part of the decline, Glasgow showed health improvements after 1850, even though immigration to the city continued unabated.

16. See, for example, Clarke (1928), pp. 7–12, and Flinn (1977), pp. 374–375.

17. This estimate is based on a simple regression where deaths in the city are predicted based on deaths in previous years. Specifically, using data from 1831 through 1882, deaths in each year are regressed against a trend term. By comparing the observed deaths to those predicted by the trend, one can arrive at an estimate of "excess deaths."

18. Cowan (1840), pp. 281–282. On cholera in Glasgow, see also "Statistics of the Malignant Cholera in Glasgow, 1848–9," *London Medical Gazette*, 1849, vol. iii, pp. 611–613; Crawford (1854–1855); Flinn (1977), pp. 371–375; and Fraser and Maver (1996), pp. 352–356.

19. Flinn (1977), p. 374.

20. Both quotations are from Clarke (1928), p. 13.

21. For a brief discussion of Snow's discovery and its broader context and significance, see Melosi (2000), p. 55.

22. See Fraser and Maver (1996), p. 4, for a brief discussion of how classical liberalism dominated Glasgow politics.

23. Fraser and Maver (1996), pp. 454–456; Clarke (1928), pp. 10–15; and Burnet (1869), pp. 43–46.

24. Clarke (1928), pp. 11–14.

25. Fraser and Maver (1996), p. 457.

26. On municipal ownership in Glasgow, see Allan (1965), Crawford (1906), and Smart (1895a, 1895b).

27. See, for example, Parsons (1904) and Crawford (1906).

28. Clarke (1928), p. 12.

29. Burnet (1869), p. 38.

30. Burnet (1869), p. 52.

31. Burnet (1869), p. 54.

32. Burnet (1869), p. 54.

33. Burnet (1869), p. 50.

34. Burnet (1869), p. 52.

35. Burnet (1869), p. 57.

36. Burnet (1869), pp. 39 and 65–67.

37. On the use of Clyde water, see Burnet (1869), p. 158; on the purchase prices of the Glasgow and Gorbals water systems, see Burnet (1869), p. 102. For other material in the paragraph, see Taylor (1899) and Clarke (1928), pp. 28–35.

38. Burnet (1869), pp. 98–110 and 116–117; Taylor (1899); and Clarke (1928), pp. 30–32. Quotations are taken from Clarke (1928), pp. 31–32. The final cost of the aqueduct is from Burnet (1869), p. 126. The historical cost has been recalculated in modern terms using the "How Much Is That Worth" calculator available at www.EH.net.

39. Crawford (1906), pp. 1–2.

40. Crawford (1906), p. 9.

41. Fraser and Maver (1996), pp. 454–455.

42. Clarke (1928), pp. 47–48.

43. Clarke (1928), p. 26.

44. Crawford (1906), p. 9, and Clarke (1928), p. 19.

45. Penny's reports are reprinted in Kirkwood (1859), pp. 212–253.

46. Kirkwood (1859), p. 254.

47. The numerous reports and experiments conducted by these chemists are reprinted in Kirkwood (1859), pp. 254–285.

48. Quotations are from Kirkwood (1859), pp. 284–285.

49. Smaller towns and boroughs mentioned by the 1854 investigators have been dropped because it is impossible to explore the subsequent histories of these places, even at a cursory level.

50. Kirkwood (1859), pp. 256–257.

51. See Smith (1852).

52. Kirkwood (1859), pp. 253–254.

53. This paragraph is based on Clarke (1928), pp. 18–20.

54. Clarke (1928), pp. 19–20.

55. Parkes and Kenwood (1901), pp. 51–52.

56. Reade (1921), p. 19.

57. Saunders (1882), p. 16.

58. Richardson (1895), p. 265.

59. See *British Medical Journal*, November 8, 1890, p. 1089. See also the discussion in chapter 6 of epidemic lead poisoning in northern England.

60. Local Government Board (1893–1894), pp. 284–285.

61. Thresh (1901), pp. 234–235. See also Thresh (1922), pp. 501–502; Richardson (1895), p. 265; and Thorne (1886).

62. Clarke (1928), pp. 28–29.

63. See Melosi (2000), pp. 83–84 and 127–129, for detailed discussion of these other projects.

64. Cholera was getting worse in Glasgow despite the best efforts of health officials in the city. According to Crawford (1854–1855), city health officials tried to

contain the 1853–1854 epidemic through quarantines and the establishment of special hospitals for cholera victims.

65. Fraser and Maver (1996), p. 157.

66. See Cutler and Miller (2004) and Troesken (2004).

67. For a broader analysis of the evolution of Glasgow's disease profile over the course of the nineteenth century, see Lees (1996).

68. See Ferrie and Troesken (2004).

69. One of the physicians who first recommended the implementation of the Loch Katrine aqueduct, Robert Dundas Thomson, wrote an essay arguing that the experience of Glasgow vindicated his recommendation. According to Thomson, experience showed that, although the water did have a tendency to dissolve some lead, it was not sufficient to cause injury. Interestingly, Thomson recommended that the city stop using lead piping anyway. See Thomson (1854).

70. The remarks of E. J. Mills were reported in many smaller newspapers across the United States. See, for example, *Bismarck Daily Tribune*, September 24, 1891, p. 4, and *Bucks County Daily Gazette* (Pennsylvania), August 20, 1891, p. 3. Strangely, larger newspapers such as the *New York Times* do not appear to have covered the story.

71. Taylor (1899). The quotation is from p. 951.

72. See the following articles: Moore et al. (1981), Moore et al. (1982), Moore et al. (1979), and Moore (1977).

73. See Jones (1989), p. 669, for a brief summary of Moore's findings regarding the health effects of lead-contaminated water. For additional evidence pointing to the same conclusion, see Crawford and Morris (1967) and an unsigned article in *Lancet*, November 18, 1967, pp. 1076–1078. See also Beattie et al. (1972), who presented evidence that water lead in Glasgow was causing symptoms consistent with lead poisoning.

74. Beattie et al. (1975).

75. Moore et al. (1981).

76. Moore et al. (1981).

77. Watt et al. (2000). One study of Glasgow women, conducted after the lime dosing began, found that variation in water-lead levels in Glasgow homes could not explain the variation in neural tube defects across the city. See MacDonell, Campbell, and Stone (2000).

78. See Elwood, St. Leger, and Morton (1976) and Lauwerys et al. (1977). For Moore's analysis and response, see Moore et al. (1979), which showed that blood lead varied with the cubed root of water lead. See also the discussion in chapter 2 on the significance of nonlinear relationships in studying the epidemiology of lead poisoning.

79. Moore et al. (1981, p. 203) wrote: "7.5% of mothers in 1977 had blood lead levels in excess of 1.5 μmol/l but only 0.4% of the equivalent group of women had blood lead levels in excess of 1.5 μmol/l in 1980."

80. See Jones (1989), p. 669, for discussion of this issue.

81. For the quotations, see Jones (1989), pp. 668–669. The original survey results are reported in Pocock (1980). Evidence of lead-contaminated water supplies inducing serious cases of lead poisoning in rural Scotland and elsewhere in Britain is reported by Beattie et al. (1972).

82. As an example, see McDonnell (1924), pp. 21–22. For an empirical analysis of the relationship between disease and municipal ownership in water, see Troesken (1999, 2001). On the forces that gave rise to municipalization, see Troesken and Geddes (2003).

83. For quotations, see Jones (1989), p. 670.

84. Galiani, Gertler, and Schargrodsky (2005).

85. On the forces behind privatization in the United States, see Troesken (2006b) and Vitale (2001). On the difficulties confronting municipalities in their efforts to meet the standards imposed by the Safe Water Drinking Act and its subsequent amendments, see Laufenberg (1998).

86. Troesken (1999).

87. See the prologue for a detailed discussion of events in New York City.

88. Data on the characteristics of cities with populations greater than 30,000 are from United States Census Bureau (1903). Data on the use of lead service lines and the ownership regimes governing local water systems are from Baker (1897).

89. See Troesken and Beeson (2003).

9 Building on the Past

1. Howard (1923), p. 208.

2. For supporting evidence, see Watt et al. (2000); Méranger, Subramanian, and Chalifoux (1981); and Jones (1989).

3. For the data in this paragraph, compare Greene (1889) and Hills (1894) to Karalekas et al. (1976) and O'Brien (1976).

4. Branquinho and Robinson (1976), Bonnefy, Huel, and Guéguen (1985), and Englert and Höring (1994). Older studies reporting similar findings include Dagnino and Badino (1968) and Chaineux (1971). See, however, Zietz et al. (2001), who find little evidence of undue lead levels in Saxony drinking water by the year 2000. See also Cirarda (1998) who finds little evidence of undue water lead in the Spanish Basque Country.

5. Bryant (2004). However, a study of Utah schoolchildren found little evidence that elevated lead levels in a particular school was associated with blood-lead levels among children attending the school. See Costa et al. (1997).

6. See, for example, Wilson (1979) and Maugh (1984).

7. Powell et al. (1995).

8. Cryptosporidium, for example, was not discovered until 1976 and much is still unknown about it. See Foss-Mollan (2001), p. 171.

9. See Villena et al. (2003), Lodder and Husman (2005), Duizer et al. (2004), and Hambridge (2001). For example, cryptosporidium is only one to two microns long, has a hard, difficult-to-penetrate outer shell, and resists the traditional chlorine, formaldehyde, and ozone treatments used by many large, urban water systems to kill waterborne pathogens. See Foss-Mollan (2001), p. 171.

10. On the more general trade-off between organic and inorganic risks in water purification, see Craun et al. (1994a), Craun et al. (1994b), and Edwards and Dubi (2004).

11. See Foss-Mollan (2001), pp. 161–174. Foss-Mollan provides an accessible chemical rationale for the trade-off between lead solvency and viral eradication. See also Edwards and Dubi (2004), who analyze the role that chlorination and chlorination by-products play in relation to water-related lead exposure.

12. See, for example, the following articles in the *Washington Post*: "Agencies Brushed Off Lead Warnings," February 29, 2004, p. A.01; "Blood and Water," March 4, 2004, p. C.01; and "D.C. Lead Issue Was Debated for Months," March 16, 2004, p. A.01. See also "Accusations in Capital on Lead Levels in Water," *New York Times*, March 6, 2004, p. A.7. For a concise overview of the events in Washington, D.C., that explains the linkage between lead solvency and new water treatment processes, see Stephen (2004).

13. For a critical review of all of the studies linking cardiovascular disease to water hardness, see Sauvant and Pepin (2002). Given the role that water chemistry plays in shaping disease outcomes, one would think that there would be clear standards regarding the chemical make up of public-drinking-water supplies, yet there are few standards, and those that exist do not appear to be based on scientific literature. See Berlyne and Yagil (1973).

14. See, for example, Steckel and Rose (2002).

Appendix A: Estimating the Effects of Lead Water Pipes on Infant and Fetal Mortality

1. In particular, all regressions are estimated using STATA (7.0) and the data are weighted according to the aweights algorithm. This weighting scheme uses weights that are inversely proportional to the variance of an observation so that the variance of the jth observation is assumed to be σ^2/w_j, where w_j is the weight of the jth observation.

2. See Meeker (1972, 1976) and Preston and Haines (1991), p. 50, for discussions of the quality of the Massachusetts data. See also the related discussion in chapter 1, which provides additional evidence on data quality.

3. Preston and Haines (1991), pp. 53–59.

4. Troesken (2004), p. 49.

5. If lead use were, in fact, correlated with some unobservable variable related to both city size and health outcomes, adding controls for city size would reduce, by a large magnitude, the correlation between lead use and infant mortality. However, when the infant death rate (the death rate for children from birth to age one) is regressed against a lead-use dummy, with no other control variables included in the regression, the coefficient on the lead dummy is 160.6 with a standard error of 36.2, and a t-statistic of 4.43. And when a battery of controls for population is added to the regression, the coefficient on the lead dummy falls somewhat, to 140.6, with a standard error of 33.7, and a t-statistic of 4.17. The controls for population include the following: the population level, the natural log of population, a dummy variable equal to one if population is between 10,000 and 30,000, a dummy variable equal to one if population is between 30,000 and 60,000, and a dummy variable equal to one if population is over 60,000.

6. Dropping these older-age death rates from the regressions does not alter the findings.

7. One might wonder about the endogeneity of the sewer-system dummies. However, previously published work demonstrates that it is proper to treat sewers as exogenous determinants of disease rates (Troesken 2002).

8. See Maddala (1991), pp. 241–243.

9. There were eight towns for which it was not possible to directly identify who paid for the service lines. Based on other information provided by Baker (1897), it is assumed that in all of these towns, the property owner paid, in part or in full, for the installation of the service line.

10. Baker (1897) provides the meters-per-main variable for 62 of the 74 cities in the sample. For the remaining 12 cities, meters-per-main has been estimated using a linear regression model relating the number of water meters to the number of fire hydrants and the number of water taps.

11. Peltzman (1971).

12. For evidence on these issues, see Troesken (1997, 1999) and Troesken and Geddes (2003).

13. There is a theoretical motivation for allowing newly installed water mains to have a different effect in large cities than in small ones. Large cities tended to have higher population densities, and therefore an additional mile of mains in a large city probably implied a much larger addition to the water system, in terms of homes newly connected, than in a small city with lower population density. See also Troesken (2002) for evidence and a more thorough discussion of this issue.

14. It is not surprising to see the hardness of local water supplies affect infant mortality rates in cities using non-lead service pipes. As explained earlier in the chapter, even without lead service lines, there was likely to have been some lead elsewhere, although in much smaller amounts, in the interior plumbing of homes.

15. Cited in references as Registrar General (1883).

16. Cited in references as Local Government Board (1888–1889). Within this source, see in particular, Appendix A, Number 13, "On the Causes of Lead Poisoning by Mr. Power."

17. A series of ecological regressions using data from Massachusetts in 1915 failed to identify consistent findings regarding the effects of lead water pipes on fertility.

Appendix B: A Statistical Supplement to *The Menace and Geography of Eclampsia*

1. See chapter 6 for the origins of water treatment processes designed to mitigate lead exposure.

2. In particular, the convulsions-related death rate includes the following: inflammation of the brain, apoplexy, softening of the brain, brain paralysis, paralysis agitons, general paralysis of the insane and insanity, chorea, convulsions laryngismus stridulus, idiopathic tetanus, and paraplegia. Of these diseases and disorders, convulsions was the dominant killer and represented about one-quarter of all deaths in this category.

3. Zymotic diseases included the following: chickenpox, measles, epidemic rash, scarlet fever, typhus, relapsing fever, influenza, whooping cough, mumps, diphtheria, cerebro-spinal fever, ill-defined fevers, enteric (typhoid) fever, cholera, diarrhea, remittent fever and ague, hydrophobia, glanders, splenic fever, cowpox, syphilis, gonorrhea, phagedena, erysipelas, septicaemia, and puerperal fever (sepsis in pregnancy).

References

Adams, Horatio. 1852. "On the Action of Water on Lead Pipes, and the Diseases Proceeding From It." *Transactions of the American Medical Association*, 5:165–236.

Alderson, James. 1839. "Notices of the Effects of Lead Upon the System." *Medico-Chirurgical Transactions* 4 (2nd series):83–94.

Allan, C. M. 1965. "The Genesis of British Urban Development with Special Reference to Glasgow." *Economic History Review* 18:598–613.

Allen, Alfred H. 1882. "The Action of Water on Lead." *Chemical News* 45:145–146.

Allen, Alfred H. 1888. "The Sheffield Water Supply and Lead-Poisoning." *Sanitary Record*, February 15, 356–357.

Anawar, H. M., J. Akai, K. M. G. Mostofa, et al. 2002. "Arsenic Poisoning in Groundwater: Health Risk and Geochemical Sources in Bangladesh." *Environment International* 27:597–604.

Aub, Joseph C., Lawrence T. Fairhall, A. S. Minot, et al. 1926. *Lead Poisoning*. Vol. VII, *Medicine Monographs*. Baltimore: Williams and Wilkins.

Aykin-Burns, N., A. Laegeler, G. Kellog, et al. 2003. "Oxidative Effects of Lead in Young and Adult Fisher 344 Rats." *Archives of Environmental Contamination and Toxicology* 44: 417–420.

Baker, George. 1786. "Further Observations on the Poison of Lead." *Medical Transactions* 2:419–443.

Baker, Moses N. 1897. *The Manual of American Waterworks*. New York: Engineering News.

Barciszewska, M. Z., E. Wysko, R. Bald, et al. 2003. "5S rRNA is a Leadzyme. A Molecular Basis for Lead Toxicity." *Journal of Biochemistry* 133:309–315.

Barker, D. J. P. 1994. *Mothers, Babies, and Disease in Later Life*. London: BMJ Publishing Group.

Beattie, A. D., J. H. Dagg, A. Goldberg, et al. 1972. "Lead Poisoning in Rural Scotland." *British Medical Journal*, May 27, pp. 488–491.

Beattie, A. D., M. R. Moore, W. T. Devenay, et al. 1972. "Environmental Lead Pollution in an Urban Soft-Water Area." *British Medical Journal*, May 27, pp. 491–493.

Beattie, A. D., M. R. Moore, A. Goldberg, et al. 1975. "Role of Chronic Low-Lead Exposure in the Aetiology of Mental Retardation." *Lancet*, March 15, pp. 589–592.

Beigel, Yitzhak, Iris Ostfeld, and Nili Schoenfeld. 1998. "A Leading Question." *New England Journal of Medicine* 339:827–830.

Bell, W. Blair, R. A. Hendry, and H. E. Annett. 1925. "The Specific Action of Lead on the Chorion Epithelium of the Rabbit, Contrasted with the Action of Copper, Thallium, and Thorium." *Journal of Obstetrics and Gynaecology of the British Empire* 32:1–16.

Bellia, Anthony J., Jr. 1992. "Lead Poisoning in Children: A Proposed Legislative Solution to Municipal Liability for Furnishing Lead-Contaminated Water." *Notre Dame Law Review* 68:399–323.

Bellinger, David C. 2004. "Lead." *Pediatrics* 113:1016–1022.

Bellinger, D., A. Leviton, M. Rabinowitz, et al. 1991. "Weight Gain and Maturity in Fetuses Exposed to Low Levels of Lead." *Environmental Research* 54:151–158.

Bengtsson, Tommy, Cameron Campbell, James Z. Lee, et al. 2004. *Life under Pressure: Mortality and Living Standards in Europe and Asia, 1700–1900.* Cambridge, Mass.: MIT Press.

Berlyne, G. M., and R. Yagil. 1973. "Chemical Drinking Water Standards, An Example of Guesswork?" *Desalination* 13:217–220.

Biolonczyk, C., H. Partsch, and A. Donner. 1989. "Lead Poisoning Caused by Long-Term Use of Diachylon Ointment." *Zeitschrift fur Hautkrankheiten* 64:1118–1120.

Billings, William R. 1898. *Some Details of Water-Works Construction.* 3rd ed. New York: McGraw.

Black, A. P., R. Knight, J. Batty, et al. 2002. "An Analysis of Maternal and Fetal Hair Lead Levels." *British Journal of Obstetrics and Gynaecology* 109:1295–1298.

Bois, Frederick Y., Thomas N. Tozer, Lauren Zeise, et al. 1989. "Application of Clearance Concepts to the Assessment of Exposure to Lead in Drinking Water." *American Journal of Public Health* 79:827–831.

Bonnefy, X., G. Huel, and R. Guéguen. 1985. "Variation of the Blood Lead Level as a Result of Lead Contamination of the Subjects Drinking Water." *Water Research* 19:1299–1303.

Boston Water Commissioners. 1848. *Report of the Water Commissioners on the Material Best Adapted for Distribution Water Pipes; and on the Most Economical Mode of Introducing Water into Private Houses.* Boston: J. H. Eastburn, City Printer.

Borja-Aburto, V. H., I. Hertz-Picciotto, M. Rojas Lopez, et al. 1999. "Blood Lead Levels Measure Prospectively and Risk of Spontaneous Abortion." *American Journal of Epidemiology* 150:590–597.

Boyd, Glen R., Prasad Shettey, Anne M. Sandvig, et al. 2004. "Pb in Tap Water Following Simulated Partial Lead Pipe Replacements." *Journal of Environmental Engineering* 130:1188–1197.

Boyd, Glen R., Neil K. Tarbert, Roger J. Oliphant, et al. 2000. "Lead Pipe Rehabilitation and Replacement Techniques for Drinking Water Service: Review of Available and Emerging Technologies." *Tunneling and Underground Space Technology* 15:13–24.

Bradbury, M. W., and R. Deane. 1993. "Permeability of the Blood-Brain Barrier to Lead." *Neurotoxicology* 14:131–136.

Bramwell, Edwin. 1931. "Some Clinical Pictures Attributable to Lead Poisoning, with Special Reference to the Neurological Manifestations of Plumbism." *British Medical Journal*, July 18, pp. 87–92.

Branquinho, C. L., and V. J. Robinson. 1976. "Some Aspects of Lead Pollution in Rio de Janeiro." *Environmental Pollution* 10:287–292.

Branson, Guy J. 1899. "Lead as an Abortifacient." *British Medical Journal*, December 2, p. 1593.

Bressler, J., K. A. Kim, T. Chakraborti, et al. 1999. "Molecular Mechanisms of Lead Neurotoxicity." *Neurochemical Research* 24:595–600.

Brewster, Ursula C., and Mark A. Perazella. 2004. "A Review of Chronic Lead Intoxication: An Unrecognized Cause of Chronic Kidney Disease." *American Journal of the Medical Sciences* 327:341–347.

Brown, John. 1889. *Clinical and Chemical Observations on Plumbism, Due to Lead-Polluted Water; with Hints on its Prevention.* Bacup, England: Tyne & Shepard.

Brown, John. 1890. "The Action of Potable Water on Lead." *British Medical Journal*, February 22, p. 420.

Brown, John C. 1989a. "Public Reform or Private Gain? The Case of Investments in Sanitary Infrastructure: Germany, 1880–1887." *Urban Studies* 26:2–12.

Brown, John C. 1989b. Reforming the Urban Environment: Sanitation, Housing, and Government Intervention in Germany, 1870–1910." *Journal of Economic History* 49:450–472.

Brown, Kenneth, and Gilbert L. Ross. 2002. "Arsenic, Drinking Water, and Health: A Position Paper of the American Council on Science and Health." *Regulatory Toxicology and Pharmacology* 36:162–174.

Brown, R. S., B. E. Hingerty, J. C. Dewan, et al. 1983. "Pb(ll)-Catalysed Cleavage of the Sugar-Phosphate Backbone of Yeast tRNAPhe—Implications for Lead Toxicity and Self Splicing RNA." *Nature* 303:543–546.

Bryant, S. D. 2004. "Lead-Contaminated Drinking Waters in the Public Schools of Philadelphia." *Journal of Toxicology and Clinical Toxicology* 47:287–294.

Bryce-Smith, D., R. R. Deshpande, J. Hughes, et al. 1977. "Lead and Cadmium Levels in Stillbirths." *Lancet*, May 28, p. 1159.

Budd, P., J. Montgomery, J. Evans, et al. 2004. "Human Lead Exposure in England from Approximately 5500 BP to the 16th Century AD." *Science of the Total Environment* 318:45–58.

Bunker, George C. 1921. "Discussion of R. S. Weston's Paper: 'Lead Poisoning by Water.'" *Journal of the New England Water Works Association* 35:126–136.

Burnet, John. 1869. *History of the Water Supply to Glasgow, from the Commencement of the Present Century*. Glasgow: Bell & Bain.

Burnham, John C. 2005. "Unraveling the Mystery of Why There Was No Childhood Lead Poisoning." *Journal of the History of Medicine and Allied Sciences* 60:445–477.

Bushnell, P. J., and R. E. Bowman. 1979. "Effects of Chronic Lead Ingestion on Social Development in Infant Rhesus Monkeys." *Neurobehavioral Toxicology* 1:207–219.

Cage, R. A. 1983. "The Standard of Living Debate: Glasgow, 1800–1850." *Journal of Economic History* 43:175–182.

Cain, Louis P., and Elyce J. Rotella. 2001. "Death and Spending: Urban Mortality and Municipal Expenditure on Sanitation." *Annales de Démographie Historique* 45:139–154.

Cambell, A., A. Becaria, D. K. Lahiri, et al. 2004. "Chronic Exposure to Aluminum in Drinking Water Increases Inflammatory Parameters Selectively in the Brain." *Journal of Neuroscience Research* 75:565–572.

Canfield, R. L., C. R. Henderson, D. A. Cory-Slechta, et al. 2003. "Intellectual Impairments in Children With Blood Lead Concentrations Below 10 μg per Deciliter." *New England Journal of Medicine* 348:1517–1526.

Cardew, P. T. 2003. "A Method for Assessing the Effect of Water Quality Changes on Plumbosolvency Using Random Daytime Sampling." *Water Research* 37:2821–2832.

Chaineux, J. 1971. Lead, Toxic Element in Drinking Water in the East of Liège Province." *Archives of the Belgian Medical Society* 29:573–580.

Chalmers, J. N. Marshall. 1940. "Observations on Non-industrial Lead Poisoning." *Glasgow Medical Journal* 16 (7th series):199–212.

Cirarda, F. B. 1998. "Lead in Drinking Water in the Greater Bilbao Area (Basque Country, Spain)." *Food Additives and Contaminants* 15:575–579.

Clarke, John S. 1928. *An Epic of Municipalisation: The Story of Glasgow's Loch Katrine Water Supply*. Glasgow: Forward Printing and Publishing.

Clement, M., R. Seux, and S. Rabarot. 2000. "A Practical Model for Estimating Total Lead Intake from Drinking Water." *Water Research* 34:1533–1542.

Committee on Service Pipes. 1917. New England Water Works Association. Report of the Committee on Service Pipes. Presented to the Association on March

14, 1917. Reprinted in the *Journal of the New England Water Works Association* 31, no. 3 (September 1917):323–389.

Commonwealth of Massachusetts. 1915. *Seventy-Fourth Annual Report on Births, Marriages, and Deaths*. Boston: Wright & Potter.

Costa, D. L. 2000. "Understanding the Twentieth-Century Decline in Chronic Conditions among Older Men." *Demography* 37:53–72.

Costa, R. A., K. L. Nutall, J. B. Schaffer, et al. 1997. "Suspected Case of Lead Poisoning in a Public School." *Annals of Clinical Laboratory Science* 27:413–417.

Cowan, Robert. 1840. "Vital Statistics of Glasgow, Illustrating the Sanatory [sic] Condition of the Population." *Journal of the Statistical Society of London* 3:257–292.

Cox, Charles R. 1964. *Operation and Control of Water Treatment Processes*. Geneva: World Health Organization.

Craun, G. F., R. J. Bull, R. M. Clark, et al. 1994a. "Balancing Chemical and Microbial Risks of Drinking Water Disinfection, Part I." *Journal of Water Supply Research and Technology—AQUA* 43:192–199.

Craun, G. F., S. Regli, R. M. Clark, et al. 1994b. "Balancing Chemical and Microbial Risks of Drinking Water Disinfection, Part II." *Journal of Water Supply Research and Technology—AQUA* 43:207–218.

Crawford, John. 1854–1855. "Observations on the Present Epidemic of Cholera in Glasgow." *Glasgow Medical Journal* 2:59–77.

Crawford, M. D., and J. N. Morris. 1967. "Lead in Drinking Water." *Lancet*, November 18, pp. 1087–1088.

Crawford, Robert. 1906. "Glasgow's Experience with Municipal Ownership and Operation: Water, Gas, Electricity and Street Railways." *Annals of the American Academy of Political and Social Science* 27:1–19.

Cullen, M. R., J. M. Robins, and B. Eskenazi. 1983. "Adult Inorganic Lead Intoxication: Presentation of 31 New Cases and a Review of Recent Advances in the Literature." *Medicine* 62:221–247.

Cutler, David M., and Grant Miller. 2004. "The Role of Public Health Improvements in Health Advances: The 20th Century United States." Working Paper no. 10511, National Bureau of Economic Research.

Dagnino, N., and R. Badino. 1968. "Polyneuropathy Due to Lead Poisoning from Water Pipes." *Sistemo Nervoso* 20:417–420.

Dana, Samuel L. 1848. *Lead Diseases: Treatise from the French of L. Tanquerel des Planches, with Notes and Additions on the Use of Lead Pipe and Its Substitutes*. Lowell: Daniel Bixby.

Davidson, C. M., N. J. Peters, A. Britton, et al. 2004. "Surface Analysis and Depth Profiling of Corrosion Products Formed in Lead Pipes Used to Supply Low Alkalinity Drinking Water." *Water Science and Technology* 49:49–54.

Davies, John-Mark, and Asit Mazumber. 2003. "Health and Environmental Policy Issues in Canada: the Role of Watershed Management in Sustaining Clean Drinking Water Quality at Surface Sources." *Journal of Environmental Management* 68:273–286.

Dawson, Earl B., Douglas R. Evans, Randall Kelly, et al. 2000. "Blood Cell Lead, Calcium, and Magnesium Levels Associated with Pregnancy-Induced Hypertension and Preeclampsia." *Biological Trace Element Research* 74:107–116.

Dawson, Earl B., Douglas R. Evans, and John Nosovitch. 1999. "Third-Trimester Amniotic Fluid Metal Levels Associated with Preeclampsia." *Archives of Environmental Health* 54:412–415.

Delves, H. T., and M. J. Campbell. 1993. "Identification and Apportionment of Sources of Lead in Human Tissue." *Environmental Geochemistry and Health* 15:75–84.

Demasi, M., C. A. Penatti, R. DeLucia, et al. 1996. "The Prooxidant Effect of 5-aminolevulinic Acid in the Brain Tissue of Rats: Implications in Neuropsychiatric Manifestations in Porphyrias." *Free Radical Biology and Medicine* 20:291–299.

DeMichelle, Stephen J. 1984. "Nutrition of Lead." *Comparative Biochemistry and Physiology* 78A:401–408.

Dietert, R. R., J. E. Lee, I. Hussain, et al. 2004. "Developmental Immunotoxicology of Lead." *Toxicology and Applied Pharmacology* 198:86–94.

Dilling, Walter J., and C. W. Healy. 1926. "Influence of Lead and the Metallic Ions of Copper, Zinc, Thorium, Beryllium and Thallium on the Germination of Frogs' Spawn and on the Growth of Tadpoles." *Annals of Applied Biology* 13:177–188.

Duizer, E., P. Bijkerk, B. Rock, et al. 2004. "Inactivation of Caliciviruses." *Applied and Environmental Microbiology* 70:4538–4543.

Durant, J. L., J. Chen, H. F. Hemond, et al. 1995. "Elevated Incidence of Childhood Leukemia in Woburn, Massachusetts: NIEHS Superfund Basic Research Program Searches for Causes." *Environmental Health Perspectives* 103 (suppl. 6):93–98.

Edwards, M., and A. Dubi. 2004. "Role of Chlorine and Chloramines in the Corrosion of Lead-Based Plumbing Materials." *Journal of the American Water Works Association* 96:572–586.

Elwood, P. C., A. S. St. Leger, and M. Morton. 1976. "Dependence of Blood-Lead on Domestic Water Lead." *Lancet*, June 12, p. 1295.

Emsley, John. 2001. *Nature's Building Blocks: An A–Z Guide to the Elements.* New York: Oxford University Press.

Englert, Norbert, and Helmut Höring. 1994. "Lead Concentration in Tap-Water and in Blood of Selected Schoolchildren in Southern Saxonia." *Toxicology Letters* 72:325–331.

Ernhart, Claire B. 1992. "A Critical Review of Low-Level Prenatal Lead Exposure in the Human: 1. Effects on the Fetus and Newborn." *Reproductive Toxicology Reviews* 6:9–19.

Ernhart, Claire B., Abraham W. Wolf, Mary J. Kennard, et al. 1986. "Intrauterine Exposure to Low Levels of Lead: The Status of the Neonate." *Archives of Environmental Health* 41:287–291.

Evans, Richard J. 1988. "Epidemics and Revolutions: Cholera in Nineteenth-Century Europe." *Past and Present* 120:123–146.

Ewbank, Douglas C., and Samuel H. Preston. 1990. "Personal Health Behavior and the Decline in Infant and Child Mortality: The United States, 1900–1930." In *What We Know about Health Transitions: The Cultural, Social, and Behavioral Determinants of Health*, ed. J. Caldwell, S. Findley, P. Caldwell, et al., 34–76. Canberra: Australian National University.

Farkas, Walter R., Ali Fischbein, Stephen Solomon, et al. 1987. "Elevated Urinary Excretion of β-Aminoisobutyric Acid and Exposure to Inorganic Lead." *Archives of Environmental Health* 42:96–99.

Fenner, E. D. 1850. "Special Report on Lead Poisoning in the City of New Orleans." *Southern Medical Reports* 2:247–280.

Ferrie, Joseph P., and Werner Troesken. 2004. "Death and the City: Chicago's Mortality Transition." Unpublished paper, Northwestern University and the University of Pittsburgh.

Flaten, Trond Peder. 2001. "Aluminum as a Risk Factor in Alzheimer's Disease, with Emphasis on Drinking Water." *Brain Research Bulletin* 55:187–196.

Flinn, Michael. 1977. *Scottish Population History: from the 17th Century to the 1930s*. Cambridge: Cambridge University Press.

Foss-Mollan, Kate. 2001. *Hard Water: Politics and Water Supply in Milwaukee, 1870–1995*. West Lafayette, Ind.: Purdue University Press.

Fowler, B. A., M. H. Whittaker, M. Lipsky, et al. 2004. "Oxidative Stress Induced by Lead, Cadmium, and Arsenic Mixtures: 30-Day, 90-Day, and 180-Day Drinking Water Studies in Rats: An Overview." *Biometals* 17:567–568.

Frankland, Percy F. 1890. "Lead Poisoning by Soft-Water Supplies." *Transactions of the Sanitary Institute* 11:235–241.

Fraser, W. Hamish, and Irene Maver. 1996. *Glasgow, Volume II: 1830 to 1912*. Manchester: Manchester University Press.

Galiani, Sebastian, Paul Gertler, and Ernesto Schargrodsky. 2005. Water for Life: The Impact of the Privatization of Water Services on Child Mortality." *Journal of Political Economy* 113:83–111.

Gallagher, Joseph P. 1885. "Is Lead as a Conduit for Water Detrimental to Health." In *Proceedings of the Third Annual Convention of the National Association of Master Plumbers of the United States of America*. Held in St. Louis, June 23–26, 1885. With an Appendix Containing the Papers Submitted by the Associations and Their Delegates. Cincinnati: Keating.

Garrett, John Henry. 1891. *The Action of Water on Lead: Being An Inquiry into the Cause and Mode of the Action and Its Prevention.* London: H. K. Lewis.

Gillette-Guyonnet, S., S. Andrieu, F. Nourhashemi, et al. 2005. "Cognitive Impairment and Composition of Drinking Water in Women: Findings of the EPIDOS Study." *American Journal of Clinical Nutrition* 81:897–902.

Glass, D. V. 1964. "Some Indicators of Differences between Urban and Rural Mortality in England and Wales and Scotland." *Population Studies* 17:263–267.

Gloag, Daphne. 1981. "Sources of Lead Pollution." *British Medical Journal* 282:41–44.

Godwin, Hillary Arnold. 2001. "The Biological Chemistry of Lead." *Current Opinion in Chemical Biology* 5:223–227.

Gowers, W. R. 1888. *Of Diseases of the Nervous System.* American ed. Philadelphia: P. Blakiston's Son.

Goyer, R. A. 1990. "Transplacental Transport of Lead." *Environmental Health Perspectives* 89:101–105.

Graham, Thomas, W. A. Miller, and A. W. Hoffman. 1851. "Chemical Report on the Supply of Water to the Metropolis." *Quarterly Journal of the Chemical Society* 3:375–413.

Greene, E. M. 1889. "Observations on the Occurrence of Lead in Boston Drinking-Water." *Boston Medical and Surgical Journal* 121:533–534.

Gregory, R. 1993. "Lead: A Source of Contamination of Tap Water." In *Corrosion and Related Aspects of Materials for Potable Water Supplies,* ed. P. McIntyre and A. D. Mercer, 21–46. London: Institute of Materials.

Grobler, S. R., F. S. Thuenissen, and L. S. Maresky. 1996. "Evidence of Undue Lead Exposure in Cape Town before the Advent of Leaded Petrol." *South African Medical Journal* 86:169–171.

Grulee, C. G. 1907. "Eclamplsia of the Mother as a Cause of Early Nephritis in the Child." *Archives of Pediatrics* 24:510–515.

Guinnane, Timothy. 1997. *The Vanishing Irish: Households, Migration, and the Rural Economy in Ireland, 1850–1914.* Princeton, N.J.: Princeton University Press.

Gulson, B. L., C. W. Jameson, K. R. Mahaffey, et al. 1997. "Pregnancy Increases Mobilization of Lead from Maternal Skeleton." *Journal of Laboratory and Clinical Medicine* 130:51–62.

Gurer-Orhan, H., H. U. Sabir, and H. Ozgünes. 2004. "Correlation Between Clinical Indicators of Lead Poisoning and Oxidative Stress Parameters in Exposed Workers." *Toxicology* 195:147–154.

Hackley, Barbara, and Anne Katz-Jacobson. 2003. "Lead Poisoning in Pregnancy: A Case Study with Implications for Midwives." *Journal of Midwifery and Women's Health* 48:30–38.

Haines, Michael R. 2001. "The Urban Mortality Transition in the United States, 1800–1940." Historical Paper no. 134, National Bureau of Economic Research.

Haines, Michael R., Lee A. Craig, and Thomas Weiss. 2003. "The Short and the Dead: Nutrition, Mortality, and the 'Antebellum Puzzle' in the United States." *Journal of Economic History* 63:382–413.

Hale, G. E. and B. C. Cantab. 1892. "The Importance of Obstruction to the Outflow of Urine as a Cause of Puerperal Eclampsia." *Lancet*, September 17, p. 662.

Halem, N. B., J. R. West, C. F. Forster, et al. 2001. "The Potential for Biofilm Development in Water Distribution Systems." *Water Research* 35:4063–4071.

Haley, V. B., and T. O. Talbot. 2004. "Seasonality and Trend in Blood Lead Levels of New York State Children." *BMC Pediatrics* 4:8.

Hall, Arthur. 1905. "The Increasing Use of Lead as an Abortifacient." *British Medical Journal*, March 18, pp. 584–586.

Hall, Arthur, and W. B. Ransom. 1906. "Plumbism from the Ingestion of Diachylon [lead plaster] as an Abortifacient." *British Medical Journal*, February 24, pp. 428–430.

Hambridge, A. 2001. "Review Efficacy of Alternative Water Treatment Techniques." *Health Estate* 55:23–25.

Hamilton, Alice. 1919. "Women in the Lead Industries." *Bulletin of the United States Bureau of Labor Statistics*, no. 253. Washington, D.C.: Government Printing Office.

Hamilton, Alice. 1925. *Industrial Poisons in the United States*. New York: MacMillan.

Han, S., W. Li, U. Jamil, et al. 1999. "Effects of Weight Loss and Exercise on the Distribution of Lead and Essential Trace Elements in Rats with Prior Lead Exposure." *Environmental Health Perspectives* 107:657–662.

Harris, Louis I. 1918. "A Clinical Study of the Frequency of Lead, Turpentine, and Benzine in Four Hundred Painters." *Archives of Internal Medicine* 22:128–156.

Hazen, Allen. 1905. "Purification of Water for Domestic Use." *Transactions of the American Society of Civil Engineers* 54:131–154.

Henriksen, Tore, and Torun Clausen. 2002. "The Fetal Origins Hypothesis: Placental Insufficiency and Inheritance versus Maternal Malnutrition in Well-Nourished Populations." *Acta Obstetricia et Gynecologica Scandinavica* 81:112–114.

Hernberg, Sven. 2000. "Lead Poisoning in Historical Perspective." *American Journal of Industrial Medicine* 38:244–254.

Hertz-Picciotto, Irva. 2000. "The Evidence That Lead Increases the Risk for Spontaneous Abortion." *American Journal of Industrial Medicine* 38:300–309.

Hiden, J. H. 1912. "The Nature of Puerperal Eclampsia: Its Similarity to Chloroform Poisoning, and Its Rational Treatment." *Charlotte Medical Journal* (North Carolina) 65:16–19.

Hill, R. S. 1911. "The Cause and Management of Puerperal Eclampsia." *Transactions of the Medical Association of the State of Alabama*, April 17–20, pp. 351–360.

Hills, William B. 1894. "On the Occurrence of Lead in City Drinking-Waters." *Boston Medical and Surgical Journal* 131:632–633.

Holt, L. E. 1923. "Lead Poisoning in Infancy." *American Journal of Diseases of Childhood* 25:229–233.

Hopwood, Jeremy D., Roger J. Davey, Merfyn O. Jones, et al. 2002. "Development of Chloropyromorphite Coatings for Lead Water Pipes." *Journal of Materials Chemistry* 12:1717–1723.

Howard, Charles D. 1923. "Lead in Drinking Water." *American Journal of Public Health* 13:207–209.

Hsu, P. C., and Y. L. Guo. 2002. "Antioxidant Nutrients and Lead Toxicity." *Toxicology* 180:33–44.

Hu, Howard. 1991. "Knowledge of Diagnosis and Reproductive History among Survivors of Childhood Plumbism." *American Journal of Public Health* 81:1070–1072.

Hu, Howard, and Mauricio Hernandez-Avila. 2002. "Lead, Bones, Women, and Pregnancy—The Poison Within?" *American Journal of Epidemiology* 156:1088–1091.

Hudak, P. F. 2004. "Boron and Selenium Contamination in South Texas Groundwater." *Journal of Environmental Science and Health* 39:2827–2834.

Ingleson, H. 1934. *The Action of Water on Lead With Special Reference to the Supply of Drinking Water.* Technical Paper no. 4, Water Pollution Research Board, Department of Scientific and Industrial Research. London: His Majesty's Stationary Office.

Inglis, James Charles. 1910. "Notes on the Sheffield Water Supply, and Statistics Relating Thereto." *Proceedings of the Institution of Civil Engineers* 181:1–84.

Irving, A. 1883. "On the Bagshot Sands as a Source of Water Supply." *Geological Magazine* (Decade II) 10:404–413.

Irving, A. 1885. Water Supply from the Bagshot and Other Strata." *Geological Magazine* (Decade III): 2:17–25.

Jarrar, B. M., and Z. N. Mahmoud. 2000. "Histochemical Demonstration of Changes in the Activity of Hepatic Phophatases Induced by Experimental Lead Poisoning in Male White Rates (*Rattus Norvegicus*)." *Toxicology and Industrial Health* 16:7–15.

Johnson, F. M. 1998. "The Genetic Effects of Environmental Lead." *Mutation Research* 410:123–140.

Jones, Robin Russell. 1989. "The Continuing Hazard of Lead in Drinking Water." *Lancet*, September 16, pp. 669–670.

Kahn, A. H., A. Khan, F. Ghani, et al. 2001. "Low-Level Lead Exposure and Blood lead Levels in Children: A Cross-Sectional Survey." *Archives of Environmental Health* 56:501.

Karalekas, Peter C., Jr., Gunther F. Craun, Arthur F. Hammonds, et al. 1976. "Lead and Other Trace Metals in Drinking Water in the Boston Metropolitan Area." *Journal of the New England Water Works Association* 90:151–171.

Kerr, J. M. Munro. 1933. *Maternal Mortality and Morbidity: A Study of Their Problems.* Baltimore, Md.: William Wood and Company.

Kim, K. A., T. Chakraborti, G. Goldstein, et al. 2002. "Exposure to Lead Elevates Induction of zif268 and Arc mRNA in Rates After Electroconvulsive Shock: The Involvement of Protein Kinase C." *Journal of Neuroscience Research* 15:268–277.

Kingsbury, George H. 1851. "Remarks Upon the Use of Lead as a Conduit or Reservoir for Water for Domestic Purposes, with Cases of Lead Colic Resulting from that Cause." *New York Journal of Medicine and Collateral Sciences* 6 (New Series):309–316.

Kirby, Oscar. 1896. "Corrosion of Lead Pipes." *Engineering Magazine* 12:125–126. This article was abstracted from a larger article that appeared in the London-based journal *Machinery*, August 15, 1896.

Kirker, Gilbert. 1890. "The Action of Potable Waters on Lead." *British Medical Journal*, January 11, pp. 71–72.

Kirkwood, James P., ed. 1859. *Collection of Reports and Opinions of Chemists in Regard to the Use of Lead Pipe for Service Pipe, in the Distribution of Water for the Supply Cities.* New York: Hosford.

Klein, M., P. Kaminsky, F. Barbe, et al. 1994. "Lead Poisoning in Pregnancy." *La Presse Médicale* 23:576–580.

Knowles, S. O., W. E. Donaldson, and J. E. Andrews. 1998. "Changes in Fatty Acid Composition of Lipids from Birds, Rodents, and Preschool Children Exposed to Lead." *Biological and Trace Element Research* 61:113–125.

Krigman, M. R. 1978. "Neuropathology of Heavy Metal Intoxication." *Environmental Health Perspectives* 26:117–120.

Kruetzmann, Henry. 1898. "Albuminuria During Pregnancy, Without Convulsions of the Mother, But With Eclampsia of the New-Born Infant." *Archives of Pediatrics* 18:673–677.

Kuhlmann, A. C., J. L. McGlothan, and T. R. Guilarte. 1997. "Developmental Lead Exposure Causes Spatial Learning Deficits in Adult Rats." *Neuroscience Letters* 233:101–104.

Kuo, H. W., T. Y. Hsiao, and J. S. Lai. 2001. "Immunological Effects of Lead-Term Lead Exposure among Taiwanese Workers." *Archives of Toxicology* 75:569–575.

Landes, David S. 1998. *The Wealth and Poverty of Nations: Why Some Are So Rich and Some So Poor.* New York: W. W. Norton.

Landrigan, P. J., P. Boffetta, and P. Apostoli. 2000. "The Reproductive Toxicity and Carcinogenicity of Lead: A Critical Review." *American Journal of Industrial Medicine* 38:231–243.

Lanphear, B. P., and K. J. Roghmann. 1996. "Pathways of Lead Exposure in Urban Children." *Environmental Research* 74:67–73.

Lanphear, B. P., K. Dietrich, P. Auinger, et al. 2000. "Cognitive Deficits Associated with Blood Lead Concentrations < 10 migrog/dL in U.S. Children and Adolescents." *Public Health Reports* 115:521–529.

Lanphear, B. P., R. Hornung, M. Ho, et al. 2002. "Environmental Lead Exposure During Early Childhood." *Journal of Pediatrics* 140:40–47.

Laufenberg, Scott D. 1998. "The Struggle of Cities to Implement the Safe Drinking Water Act in the Context of Intergovernmental Relations." *Drake Journal of Agricultural Law* 3:495–417.

Lauwerys, R., G. Hubermont, J. P. Buchet, et al. 1977. "Domestic Water and Lead Exposure During Pregnancy." *Lancet*, December 3, p. 1188.

Lees, R. E. 1996. "Epidemic Disease in Glasgow during the 19th Century." *Scottish Medical Journal* 41:24–27.

Legge, Thomas M., and Kenneth W. Goadby. 1912. *Lead Poisoning and Lead Absorption.* London: Edward Arnold.

Leser, C. E. V. 1955. "Variations in Mortality and Life Expectation." *Population Studies* 9:67–71.

Liang, C., R. Ji, Y. Jiang, et al. 2003. "Relations Between Trace Elements in Drinking Water and Elderly Residents' Cognitive Function." *Journal of Hygiene Research* 32:562–567.

Lidsky, T. I., and J. S. Schneider. 2003. "Lead Neurotoxicity in Children: Basic Mechanisms and Clinical Correlates." *Brain* 126 (Part I):5–19.

Lindsay, Lauder. 1859. "On the Action of Hard Waters upon Lead." *Edinburgh New Philosophical Journal* 9:245–258; 10:8–25.

Linenthal, Harry. 1914. "The Early Diagnosis of Lead Poisoning." *Journal of the American Medical Association* 62:1796–1799.

Local Government Board. 1888–1889. *Eighteenth Annual Report. Supplement Containing the Report of the Medical Officer for 1888.* London: Her Majesty's Stationary Office.

Local Government Board. 1893–1994. *Twenty-Third Annual Report. Supplement Containing the Report of the Medical Officer for 1893.* London: Her Majesty's Stationary Office.

Local Government Board. 1908–1909. *Thirty-Eighth Annual Report. Supplement Containing the Report of the Medical Officer for 1908–90.* London: Her Majesty's Stationary Office.

Local Government Board. 1913–1914. *Forty-Third Annual Report. Supplement Containing the Report of the Medical Officer for 1913–14.* London: Her Majesty's Stationary Office.

Lodder, W. J., A. M. de Roda Husman. 2005. "Presence of Noroviruses and Other Enteric Viruses in Sewage and Surface Waters in the Netherlands." *Applied and Environmental Microbiology* 71:1453–1461.

Lowndes, J. 1936. "Modern Methods in the Diagnosis of Plumbism." *Saint Thomas Medical Reports*, 44–50.

Lyhus, Randy. 1989. "Calcium Protects Against Lead Poisoning." *Environment* 31:22–26.

MacDonnell, J. E., H. Campbell, and D. H. Stone. 2000. "Lead Levels in Domestic Water Supplies and Neural Tube Defects in Glasgow." *Archives of Diseases in Childhood* 82:50–53.

MacMorran, Alexander, and Joshua Scholefield. 1914. *The Public Health Acts, Annotated with Appendices.* Vol. 1. 8th ed. London: Butterworth & Co., and Shaw & Sons.

Maddala, G. S. 1991. *Limited-Dependent and Qualitative Variables in Econometrics.* Cambridge: Cambridge University Press.

Maine State Board of Health. 1915. *Annual Report of the State Board of Health of the State of Maine.* Augusta, Me.: Burleigh & Flynt, Printers.

Markovac, J., and G. W. Goldstein. 1988. "Picomolar Concentrations of Lead Stimulate Brain Protein Kinase C." *Nature* 334:71–73.

Marsh, Leonard S. M. 1910. "Notes on the Sheffield Water-Supply, Statistics Relating Thereto." *Minutes of the Proceedings [Transactions] of the Institution of Civil Engineers* 181:1–79.

Massachusetts State Board of Health. 1871. *Second Annual Report of the State Board of Health of Massachusetts.* Public Document No. 37. Boston: Wright & Potter.

Massachusetts State Board of Health. 1878. *Ninth Annual Report of the State Board of Health of Massachusetts.* Public Document No. 30. Boston: Wright & Potter.

Massachusetts State Board of Health. 1895. *Twenty-Sixth Annual Report of the State Board of Health of Massachusetts.* Public Document No. 34. Boston: Wright & Potter.

Massachusetts State Board of Health. 1896. *Twenty-Seventh Annual Report of the State Board of Health of Massachusetts.* Public Document No. 34. Boston: Wright & Potter.

Massachusetts State Board of Health. 1898. *Twenty-Ninth Annual Report of the State Board of Health of Massachusetts.* Public Document No. 34. Boston: Wright & Potter.

Massachusetts State Board of Health. 1899. *Thirtieth Annual Report of the State Board of Health of Massachusetts.* Public Document No. 34. Boston: Wright & Potter.

Massachusetts State Board of Health. 1900. *Thirty-First Annual Report of the State Board of Health of Massachusetts*. Public Document No. 34. Boston: Wright & Potter.

Massachusetts State Board of Health. 1905. *Thirty-Sixth Annual Report of the State Board of Health of Massachusetts*. Public Document No. 34. Boston: Wright & Potter.

Maugh, Thomas H. 1984. "Acid Rain's Effects on People Assessed." *Science* 226:1408–1410.

McDonald, Jill A., and Nancy Upp Potter. 1996. "Lead's Legacy? Early and Late Mortality of 454 Lead-Poisoned Children." *Archives of Environmental Health* 51:116–121.

McDonnell, R. E. 1924. *Rates, Revenues and Results of Municipal Ownership of Water Works in the U.S.* Kansas City, Mo.: Burns & McDonnell Engineering Company.

McKay, Jennifer, and Anthony Moeller. 2002. "Are Mandatory Regulations Required for Water Quality in Australia." *Water Policy* 4:95–118.

McKeown, Thomas. 1976. *The Modern Rise of Population*. New York: Academic Press.

McNeill, William H. 1977. *Plagues and Peoples*. New York: Anchor Books.

Meeker, Edward. 1972. "The Improving Health of the United States, 1850–1915." *Explorations in Economic History* 9:353–373.

Meeker, Edward. 1976. "The Social Rate of Return on Investment in Public Health, 1880–1910." *Journal of Economic History* 34:392–421.

Melosi, Martin V. 2000. *The Sanitary City: Urban Infrastructure in America from Colonial Times to the Present*. Baltimore, Md.: Johns Hopkins University Press.

Méranger, J. C., K. S. Subramanian, and C. Chalifoux. 1981. "Survey for Cadmium, Cobalt, Chromium, Copper, Nickel, Lead, Zinc, Calcium, and Magnesium in Canadian Drinking Water Supplies." *Journal of the Association of Official Analytical Chemists* 64:44–53.

Mesch, U., R. M. Lowenthal, and D. Coleman. 1996. "Lead Poisoning Masquerading as Chronic Fatigue Syndrome." *Lancet* 347:1193.

Meyer, I., J. Heinrich, M. J. Trepka, et al. 1998. "The Effect of Lead in Tap Water on Blood Lead in Children in a Smelter Town." *Science of the Total Environment* 209:255–271.

Milligan, Ernest. 1931. "Correspondence." *British Medical Journal*, August 1, pp. 222–223.

Mitsui, K. 1934. "An Experiment of the Lead-Poisoning in Pregnancy." *Japanese Journal of Obstetrics* 17:304–308.

Mokyr, Joel. 2000. "Why 'More Work for Mother?' Knowledge and Household Behavior, 1870–1945." *Journal of Economic History* 60:1–41.

Moore, Michael R. 1977. "Lead in Drinking Water in Soft Water Areas—Health Hazards." *Science of the Total Environment* 7:109–115.

Moore, Michael R., Abraham Goldberg, Peter A. Meredith, et al. 1979. *Clinica Chimica Acta* 95:129–133.

Moore, Michael R., Abraham Goldberg, W. Morton Frye, and W. N. Richards. 1981. "Maternal Lead Levels After Alterations to Water Supply." *Lancet*, July 25, pp. 203–204.

Moore, M. R., A. Goldberg, S. J. Pocock, et al. 1982. "Some Studies of Maternal and Infant Lead Exposure in Glasgow." *Scottish Medical Journal* 27:113–122.

Moorhouse, S. R., S. Carden, P. N. Drewitt, et al. 1988. "The Effect of Chronic Low Level Lead Exposure on Blood-Brain Barrier Function in the Developing Rat." *Biochemical Pharmacology* 37:4539–4547.

Murray, John. 1997. "Standards of the Present for People of the Past: Height, Weight, and Mortality among Men of Amherst College." *Journal of Economic History* 57:585–606.

Murray, Leith. 1911. "Puerperal Eclampsia: A Comparison With Venom Poisoning and a Suggestion for Treatment Following Therefrom." *British Medical Journal*, January 28, pp. 185–187.

Myers, C. N., Florence Gustafson, and Binford Throne. 1935. "The Distribution and Diagnostic Significance of Lead in the Human Body." *New York State Journal of Medicine* 35:579–583.

Nash, Gifford. 1892. "The Importance of Obstruction to the Outflow of Urine as a Cause of Puerperal Eclampsia." *Lancet*, August 27, pp. 477–480.

Nation, J. R., D. M. Baker, B. Taylor, et al. 1986. "Dietary Lead Increases Ethanol Consumption in the Rat." *Behavioral Neuroscience* 100:525–530.

Needleman, Herbert L. 1997. "Clamped in a Straightjacket: The Insertion of Lead into Gasoline." *Environmental Research* 74:95–103.

Needleman, Herbert L. 1998. "Clair Patterson and Robert Kehoe: Two Views on Lead Toxicity." *Environmental Research* 78:79–85.

Needleman, Herbert L. 2000. "The Removal of Lead from Gasoline: Historical and Personal Reflections." *Environmental Research* 84:20–25.

Needleman, Herbert L. 2004. "Lead Poisoning." *Annual Review of Medicine* 55:209–222.

Needleman, Herbert L., and Constantine A. Gastonis. 1990. "Low-Level Lead Exposure and the IQ of Children: A Meta-analysis of Modern Studies." *Journal of the American Medical Association* 263:673–678.

Needleman, Herbert L., Christine McFarland, Roberta B. Ness, et al. 2002. "Bone Lead Levels in Adjudicated Delinquents: A Case Control Study." *Neurotoxicology and Teratology* 24:711–717.

Needleman, Herbert L., Michael Rabinowitz, Michael Burley, et al. 1984. "Lead in Umbilical Blood, Indoor Air, Tap Water, and Gasoline in Boston." *Archives of Environmental Health* 39:299–301.

Needleman, Herbert L., Alan Schell, David Bellinger, et al. 1990. "The Long-Term Effects of Exposure to Low Doses of Lead in Childhood: An Eleven-Year Follow-Up Report." *New England Journal of Medicine* 322:83–88.

Nehru, B., and S. Kaushal. 1993. "Alterations in the Hepatic Enzymes Following Experimental Lead Poisoning." *Biological Trace Element Research* 38:27–34.

New York City. 1869. *Fourth Annual Report of the Metropolitan Board of Health of the State of New York, 1869.* New York: D. Appleton.

Newton, A. C. 1995. "Protein Kinase C: Structure, Function, and Regulation." *Journal of Biological Chemistry* 270:495–498.

Nichols, James R. 1860. "Local Decomposition in Lead Aqueduct Pipes." *Boston Medical and Surgical Journal* 63:149–152.

Nriagu, Jerome O. 1983. *Lead and Lead Poisoning in Antiquity.* New York: Wiley & Sons.

Nriagu, Jerome O. 1998. "Tales Told in Lead." *Science* 281:1622–1623.

O'Brien, Joseph E. 1976. "Lead in Boston Water: Its Cause and Prevention." *Journal of the New England Water Works Association* 91:173–180.

O'Gráda, Cormac. 1999. *Black '47 and Beyond: The Great Irish Famine in History, Economy, and Memory.* Princeton, N.J.: Princeton University Press.

Oktem, F., M. K. Arslan, B. Dündar, et al. 2004. "Renal Effects and Erythrocyte Oxidative Stress in Long-Term Low-Level Lead-Exposed Adolescent Workers in Auto Repair Workshops." *Archives of Toxicology* 12:681–687.

Oliveira, Steve, Aro Antonio, David Sparrow, et al. 2002. "Season Modifies the Relationship between Bone Lead and Blood Lead Levels: The Normative Aging Study." *Archives of Environmental Health* 57:466–472.

Oliver, Thomas. 1897. "Metallic and Some Other Forms of Poisoning: Including Poisonous Trades." In *A System of Medicine*, vol. 3, ed. Thomas Clifford Albutt. New York: Macmillan.

Oliver, Thomas. 1911. "Lead Poisoning and the Race." *Eugenics Review* 3:83–93.

Oliver, Thomas. 1914. *Lead Poisoning: From the Industrial, Medical, and Social Points of View.* London: H. K. Lewis.

Onalaja, Ava O., and Luz Claudio. 2000. "Genetic Susceptibility to Lead Poisoning." *Environmental Health Perspectives* 108:23–28.

Page, N. M. 2002. "The Endocrinology of Preeclampsia." *Clinical Endocrinology* 57:413–423.

Papaioannou, N., I. Vlemmas, N. Balaskas, et al. 1998. "Histopathological Lesions in Lead Intoxicated Dogs." *Veterinary and Human Toxicology* 40:203–207.

Parkes, Louis and Henry Kenwood. 1901. *Hygiene and Public Health.* Philadelphia: P. Blakiston's Son.

Parsons, Frank. 1904. "Glasgow's Great Record: A Complete History of the Pioneer Experiment in Municipal Ownership of Street-Car Service in Great Britain." *The Arena* 32:461–471.

Paul, Bimal Kanti. 2004. "Arsenic Contamination Awareness Among the Rural Residents in Bangladesh." *Social Science and Medicine* 59:1741–1755.

Peckham, C. H. 1935. "An Analysis of 127 Cases of Eclampsia Treated by the Modified Stroganoff Method." *American Journal of Obstetrics and Gynecology* 29:27–35.

Peltzman, Sam. 1971. "Pricing at Public and Private Enterprises: Electric Utilities in the United States." *Journal of Law and Economics* 14:109–147.

Peplow, E., and R. Edmonds. 2004. "Health Risks Associated with Contamination of Groundwater by Abandoned Mines Near Twisp in Okanogan County, Washington." *Environmental Geochemistry and Health* 26:69–79.

Perazella, M. A. 1996. "Lead and the Kidney: Nephropathy, Hypertension, and Gout." *Connecticut Medicine* 60:521–526.

Perleman, S., L. Hertz-Pannier, M. Hassan, et al. 1993. "Lead Encephalopathy Mimicking a Cerebellar Tumor." *Acta Paediatrica* 82:423–425.

Pocock, Stuart J. 1980. "Factors Influencing Household Water Lead: A British National Survey." *Archives of Environmental Health* 35:45–51.

Pocock, Stuart J., Marjorie Smith, and Peter Baghurst. 1994. "Environmental Lead and Children's Intelligence: A Systematic Review of the Epidemiological Evidence." *British Medical Journal* 309:1189–1197.

Poincaré, Henri. 1952. *Science and Hypothesis*. New York: Dover Publications.

Pope, Frank M. 1893. "Two Cases of Poisoning by the Self-Administration of Diachylon—Lead Plaster—For the Purpose of Procuring Abortion." *British Medical Journal*, July 1, pp. 9–10.

Porritt, Norman. 1883. *The Operative Treatment of Intra-Thoracic Effusion*. London: J. and A. Chruchill.

Porritt, Norman. 1891. *Cornered*. London: Leadenhall Press.

Porritt, Norman. 1895. *Religion and Health: Their Mutual Relationship and Influence*. London: Skeffington.

Porritt, Norman. 1916. *The Drug Danger: A Warning*. London.

Porritt, Norman. 1923. *The Factory King*. London: J. Cape.

Porritt, Norman. 1926. *The Abdomen in Labor: Being A General Practitioner's Clinical Study of the Parturient Abdomen*. London: Oxford University Press, Oxford Medical Publications.

Porritt, Norman. 1931. "Cumulative Effects of Infinitesimal Doses of Lead." *British Medical Journal*, July 18, pp. 92–94.

Porritt, Norman. 1934. *The Menace and Geography of Eclampsia in England and Wales*. London: Humphrey Milford and Oxford University Press.

Powell, J. J., S. M. Greenfield, R. P. Thompson, et al. 1995. "Assessment of Toxic Metal Exposure Following the Camelford Water Pollution Incident: Evidence of Acute Mobilization of Lead Into Drinking Water." *Analyst* 120:793–798.

Preston, Samuel H., and Michael R. Haines. 1991. *Fatal Years: Child Mortality in Late-Nineteenth-Century America*. Princeton, N.J.: Princeton University Press.

Preston, Samuel H., and Etienne van de Walle. 1978. "Urban French Mortality in the Nineteenth Century." *Population Studies* 32:275–297.

Puklová, V., A. Batáriová, M. Cervá, et al. 2005. "Cadmium Exposure Pathways in the Czech Urban Population." *Central European Journal of Public Health* 13:11–19.

Putnam, James J. 1887. "On the Frequency With Which Lead Is Found in the Urine, and on Certain Points in the Symptomatology of Chronic Lead Poisoning." *Boston Medical and Surgical Journal*, July 28, pp. 73–76, and August 4, pp. 97–99.

Putnam, James J. 1889. "A Supplemental Inquiry into the Frequency With Which Lead Is Found in the Urine." *Boston Medical and Surgical Journal*, November 28, pp. 530–533.

Quam, G. N. and Arthur Klein. 1936. "Lead Water Pipes as a Source of Lead in Drinking Water." *American Journal of Public Health* 26:778–780.

Ransom, W. B. 1900. "On Lead Encephalopathy and the Use of Diachylon [lead plaster] as an Abortifacient." *British Medical Journal*, June 30, pp. 1590–1592.

Ravin, J. G., and T. B. Ravin. 1999. "What Ailed Goya?" *Journal of Ophthalmology* 44:163–170.

Rea, O. A. 1894–1895. "Some of My Experience With Puerperal Eclampsia—Conclusions as to Its Cause and Treatment." *Louisville Medical Monthly* 1:424–429.

Reade, James F. 1921. "Lead Poisoning and Water Supplies." *Transactions of the Institution of Civil Engineers of Ireland* 47:12–38.

Rees, O. W., and A. L. Elder. 1928. "The Effect of Certain Illinois Waters on Lead." *Journal of the American Water Works Association* 15:714–724.

Registrar General. 1870. *Thirty-Third Annual Report of Births, Deaths, and Marriages in England*. London: Eyre and Spottsiwoode.

Registrar General. 1883. *Forty-Sixth Annual Report of Births, Deaths, and Marriages in England*. London: Eyre and Spottsiwoode.

Reyes, Jessica Wolpaw. 2005. "The Impact of Prenatal Lead Exposure on Health." Unpublished manuscript, Department of Economics, Amherst College.

Rice, D. C. 1992. "Behavioral Impairment Produced by Developmental Lead Exposure: Evidence from Primate Research." In *Human Lead Exposure*, ed. Hebert L. Needleman, 137–154. Boca Raton, Fla.: CRC Publisher.

Richardson, Sir Benjamin Ward. 1895. "Note on the Position of the Lead and Water Question." *Asclepiad* 11:264–268.

Rosenberg, Charles E. 1971. *The Cholera Years: The United States in 1832, 1849, and 1866.* Chicago: University of Chicago Press.

Rothenberg, S. J., V. Kondrashov, M. Manalo, et al. 2002. "Increases in Hypertension and Blood Pressure During Pregnancy with Increased Bone Lead Levels." *American Journal of Epidemiology* 156:1079–1087.

The Sanitarian. 1899. Vol. XLIII, July–December.

Sartor, F., and D. Rondia. 1980. "Blood Lead Levels and Age: A Study of Two Male Populations not Occupationally Exposed." *Archives of Environmental Health* 35:110–116.

Sata, R., S. Araki, T. Tanigawa, et al. 1998. "Changes in T Cell Subpopulations in Lead Workers." *Environmental Research* 67:168–182.

Satarug, S., M. Nishijo, P. Ujjin, et al. 2004. "Evidence for Concurrent Effects of Exposure to Environmental Cadmium and Lead on Hepatic CYP2A6 Phenotype and Renal Function Biomarkers in Nonsmokers." *Environmental Health Perspectives* 112:1512–1518.

Sauer, R. 1978. "Infanticide and Abortion in Nineteenth-Century Britain." *Population Studies* 32:81–93.

Saunders, W. Sedgwick. 1882. "On the Action of Water Upon Lead Pipes, Being a Translation from the French of M. Belgrand." *Chemical News* 65:7–9, 14–16.

Sauvant, M. P., and D. Pepin. 2002. "Drinking Water and Cardiovascular Disease." *Food and Chemical Toxicology* 40:1311–1125.

Schaut, George C. 1942. "The Action of Chlorinated Water Supply Upon Lead Pipe." *American Journal of Pharmacy* 114:241–249.

Schettler, Ted, Gina Solomon, Maria Valenti, et al. 2000. *Generations at Risk: Reproductive Health and the Environment.* Cambridge, Mass.: MIT Press.

Schlembach, D. 2003. "Pre-Eclampsia—Still a Disease of Theories." *Fukushima Journal of Medical Science* 49:69–115.

Schneider, J. S., M. H. Lee, D. W. Anderson, et al. 2001. "Enriched Environment during Development is Protective Against Lead-Induced Neurotoxicity." *Brain Research* 896:48–55.

Schütz, A., L. Barregård, G. Sällsten, et al. 1997. "Blood Lead in Uruguayan Children and Possible Sources of Exposure." *Environmental Research* 74:17–23.

Schwartz, J. 1991. "Lead, Blood Pressure, and Cardiovascular Disease in Men and Women." *Environmental Health Perspectives* 91:71–75.

Schwartz, J. 1993. "Beyond LOEL's, p Values, and Vote Counting: Methods for Looking at the Shapes and Strengths of Correlations." *Neurotoxicology* 14:237–246.

Sedgwick, W. T. 1901. "On the Rise and Progress of Water-Supply Sanitation in the Nineteenth Century." *Journal of the New England Water Works Association* 15:315–337.

Sen, Amartya. 1981. *Poverty and Famines.* Oxford: Clarendon Press.

Shannon, Michael W., and John W. Graef. 1992. "Lead Intoxication in Infancy." *Pediatrics* 89:87–90.

Shapiro, Irving M., Bruce Dobkin, Orhan C. Tuncay, et al. 1973. "Lead Levels in the Dentine and Cicumpulpal Dentine of Deciduous Teeth of Normal and Lead Poisoned Children." *Clinica Chimica Acta* 46:119–123.

Sharp, David W. A. 2003. *The Penguin Dictionary of Chemistry*. New York: Penguin Books.

Simmons, T. 1993. "Lead-Calcium Interactions in Cellular Lead Toxicity." *Neurotoxicology* 14:77–85.

Smart, William. 1895a. "Glasgow and its Municipal Industries." *Quarterly Journal of Economics* 9:188–194.

Smart, William. 1895b. "The Municipal Work and Finance of Glasgow." *Economic Journal* 5:35–49.

Smith, C. Chlemesha. 1910. "Methods of Treating Water to Prevent Plumbo-Solvency." *Journal of the Royal Institute for Public Health* 18:83–90.

Smith, C. Chlemesha, and E. M. Chaplin. 1904. "The Treatment of Moorland Water to Prevent Action upon Lead." *Transactions of the British Association of Water Engineers* 9:184–207.

Smith, Frederick L., Thomas K. Rathmell, and George E. Marcil. 1938. "The Early Diagnosis of Acute and Latent Plumbism." *American Journal of Clinical Pathology* 8:471–508.

Smith, John. 1852. "On the Composition of the Waters of the Dee and Don, at Aberdeen, with an Investigation into the Action of Dee Water on Lead Pipes and Cisterns." *The Quarterly Journal of the Chemical Society* 4:123–133.

Smitherman, J., and P. Harber. 1991. "A Case of Mistaken Identity: Herbal Medicine as a Case of Lead Toxicity." *American Journal of Industrial Medicine* 20:795–798.

Solliway, Bernard M., Alex Schaffer, Hillel Pratt, et al. 1994. "A Multidisciplinary Study of Lead Exposed Subjects: 1. Delayed Target Detection P-300 Latency, an Electrophysiological Parameter, Correlates with Urinary δ-ALA." *Environmental Research* 67:168–182.

Song, Chen, and Louis L. Nguyen. 2003. "The Effect of Hernias on the Labor Force Participation of Union Army Veterans." In *Health and Labor Force Participation over the Life Cycle: Evidence from the Past*, ed. Dora L. Costa, 253–310. Chicago: University of Chicago Press.

Sowers, Maryfran, Mary Jannausch, Theresa Scholl, et al. 2002. "Blood Lead Concentrations and Pregnancy Outcomes." *Archives of Environmental Health* 57:489–495.

Stainthorpe, W. W. 1914. "Observations of 120 Cases of Lead Absorption from Drinking-Water." *Lancet*, July 25, pp. 213–215.

Steckel, Richard H. 1995. "Stature and the Standard of Living." *Journal of Economic Literature* 33:1903–1940.

Steckel, Richard H., and Jerome C. Rose, eds. 2002. *The Backbone of History: Health and Nutrition in the Western Hemisphere.* New York: Cambridge University Press.

Stephen, Andrew. 2004. "In the Land Where Lives Are Health to Be More Precious than Anywhere Else, the Capital's Denizen's Are Warned that They May Be Getting Contaminated Water from Their Taps." *New Statesman*, March 22, pp. 17–18.

Stewart, D. D. 1893. *Treatment of Typhoid Fever.* Detroit: George S. Davis.

Strang, John. 1859. "On Water Supply to Great Towns: Its Extent, Cost, Uses, and Abuses." *Journal of the Statistical Society of London* 22:232–249.

Suszkiw, J. B. 2004. "Presynaptic Disruption of Transmitter Release by Lead." *Neurotoxicology* 25:599–604.

Swann, Alfred. 1889. "On Lead Poisoning from Service Pipes, in Relation to Sterility and Abortion." *British Medical Journal*, February 16, p. 352.

Swann, Alfred. 1892. "A National Danger: Lead Poisoning from Service Pipes." *Lancet*, July 23, pp. 194–195.

Sylvester, John E. 1899–1900. "Puerperal Eclampsia: Its Cause and Treatment." *Annals of Gynecology and Pediatry* 13:740–749.

Szreter, Simon. 1988. "The Importance of Social Intervention in Britain's Mortality Decline, c. 1850–1914." *Social History of Medicine* 1:1–37.

Szreter, Simon. 1997. "Economic Growth, Disruption, Deprivation, Disease, and Death: On the Importance of the Politics of Public Health for Development." *Population and Development Review* 23:693–728.

Szreter, Simon, and Graham Mooney. 1998. "Urbanization, Mortality, and the Standard of Living Debate: New Estimates of the Expectation of Life at Birth in Nineteenth-Century British Cities." *Economic History Review* 49:84–112.

Tattersall, C. H. 1897. "The Action of Moorland Water on Lead and the Prevention of the Same." *Journal of the Sanitary Institute* 18:625–634.

Taylor, Benjamin. 1899. "The Water Supply of the City of Glasgow." *Engineering Magazine* 17:937–952.

Thain, M., and M. Hickman. 2000. *The Penguin Dictionary of Biology.* 10th ed. New York: Penguin.

Theobald, G. W. 1930. "The Causation of Eclampsia: Observations and Experiments." *Lancet*, May 24, pp. 1115–1123.

Thomson, G. O., G. M. Raab, W. S. Hunter, et al. 1989. "Blood-Lead Levels and Children's Behavior—Results from the Edinburgh Lead Study." *Journal of Child Psychology and Psychiatry, and Allied Disciplines* 30:515–528.

Thomson, John. 1923. "Peculiar Fatal Convulsions in Four Children, Whose Father Suffered from Lead Poisoning." *British Journal of Children's Diseases* 20:193–196.

Thomson, Robert Dundas. 1854. "Lead Cisterns and Pipes in a Sanitary Point of View." *Lancet*, ii, pp. 79–80.

Thomson, William. 1882. "Notes on Lead Pipes and Lead Contamination." *Chemical News* 45:116–118.

Thorne, R. 1886. "Lead-Poisoning by Potable Water." *Practitioner* 37:464–480.

Thresh, John C. 1901. *Water and Water Supplies*. 3rd ed. Philadelphia: P. Blakiston's Son.

Thresh, John C. 1905. "A Series of Cases of Lead Poisoning Due to Hard Water." *Lancet*, October 7, pp. 1033–1034.

Thresh, John C. 1922. "The Action of Natural Waters on Lead, Parts I and II." *Analyst* 47:459–468, 501–505.

Thresh, John C., and John F. Beale. 1925. *The Examination of Waters and Water Supplies*. 3rd ed. Philadelphia: P. Blakiston's Son.

Todorovic, T., and D. Vujanovic. 2002. "The Influence of Magnesium on the Activity of Some Enzymes (AST, ALT, ALP) and Lead Content in Some Tissues." *Magnesium Research* 15:173–177.

Tong, Shilu, Yasmin E. von Schirnding, and Tippawan Prapamontol. 2000. "Environmental Lead Exposure: A Public Health Problem of Global Dimensions." *Bulletin of the World Health Organization* 78:1068–1077.

Tourmaa, Tuula E. 1995. "The Adverse Effects of Lead." *Journal of Orthomolecular Medicine* 10:149–164.

Townsend, M. W. 1865–1866. "Eclampsia—Poisoning by Opium." *Buffalo Medical and Surgical Journal* 5:316–319.

Troesken, Werner. 1997. "The Sources of Public Ownership: Historical Evidence from the Gas Industry." *Journal of Law, Economics, and Organization* 13:1–27.

Troesken, Werner. 1999. "Typhoid Rates and the Public Acquisition of Private Waterworks, 1880–1925." *Journal of Economic History* 59:927–948.

Troesken, Werner. 2001. "Race, Disease, and the Provision of Water in American Cities, 1889–1925." *Journal of Economic History* 61:750–777.

Troesken, Werner. 2002. "The Limits of Jim Crow: Race and the Provision of Water and Sewerage in American Cities, 1880–1925." *Journal of Economic History* 62:734–773.

Troesken, Werner. 2004. *Water, Race, and Disease*. Cambridge, Mass.: MIT Press.

Troesken, Werner. 2006a. "Lead Exposure and Eclampsia in Britain, 1883–1934." *Environmental Research*, forthcoming.

Troesken, Werner. 2006b. "Regime Change and Corruption: A History of Public Utility Regulation." In *Corruption and Reform: Lessons from America's Economic History*, ed. Claudia Goldin and Edward Glaeser, 259–284. Chicago: University of Chicago Press.

Troesken, Werner, and Patricia E. Beeson. 2003. "The Significance of Lead Water Mains in American Cities: Some Historical Evidence." In *Health and Labor Force Participation over the Life Cycle: Evidence from the Past*, ed. Dora L. Costa, 181–202. Chicago: University of Chicago Press.

Troesken, Werner, and Rick Geddes. 2003. "Municipalizing American Water-works, 1897–1917." *Journal of Law, Economics, and Organization* 8:187–206.

Tweedy, Hastings. 1919. "The Cause of Eclampsia." *Dublin Journal of Medical Science* 148:225–229.

United States Environmental Protection Agency. 1995. *Seasonal Rhythms of Blood Lead Levels: Boston, 1979–1983*. Office of Prevention, Pesticides, and Toxic Substances. Available online at www.epa.gov/lead/season.pdf.

van Der Leer, D., N. P. Weatherhill, R. J. Sharp, et al. 2002. "Modelling the Diffusion of Lead into Drinking Water." *Applied Mathematical Modelling* 26:681–699.

van Poppel, F., and C. van der Heijden. 1997. "The Effects of Water Supply on Infant and Childhood Mortality: A Review of Historical Evidence." *Health Transition Review: The Cultural, Social, and Behavioral Determinants of Health* 7:113–148.

Villena, C., R. Gabrieli, R. M. Pintó, et al. 2003. "A Large Infantile Gastroenteritis Outbreak in Albania Caused by Multiple Emerging Rotavirus Genotypes." *Epidemiology and Infection* 131:1105–1110.

Vincent, Marco, Grazia Naccia, Enrico Rocchi, et al. 2000. "Mortality in a Population with Long-Term Exposure to Inorganic Selenium via Drinking Water." *Journal of Clinical Epidemiology* 53:1062–1068.

Vinovskis, M. 1972. "Mortality Rates and Trends in Massachusetts before 1860." *Journal of Economic History* 32:184–213.

Vinovskis, M. 1981. *Fertility in Massachusetts from the Revolution to the Civil War*. New York: Academic Press.

Vitale, Robert. 2001. "Privatizing Water Systems: A Primer." *Fordham International Law Journal* 24:1382–1404.

Waldron, H. A. 1973. "Lead Poisoning in the Ancient World." *Medical History* 17:391–399.

Warren, Christian. 2000. *Brush with Death: A Social History of Lead Poisoning*. Baltimore, Md.: Johns Hopkins University Press.

Waters, Ernest E. 1894. "A Case of Puerperal Eclampsia Following Lead Poisoning." *British Medical Journal*, March 31, pp. 682–683.

Watt, G. C. M., A. Britton, H. G. Gilmour, et al. 2000. "Public Health Implications of New Guidelines for Lead in Drinking Water: A Case-Study in an Area with Historically High Water Lead Levels." *Food and Chemical Toxicology* 38(suppl.):S73–S79.

Wedeen, Richard P. 1984. *Poison in the Pot: The Legacy of Lead*. Carbondale: Southern Illinois University Press.

Weston, Robert Spurr. 1920. "Lead Poisoning by Water and Its Prevention." *Journal of the New England Water Works Association* 34:239–263.

Whipple, George C. 1913. "Decarbonation as a Means of Removing the Corrosive Properties of Public Water Supplies." *Journal of the New England Water Works Association* 27:193–227.

White, Sinclair. 1889. "A Discussion on the Contamination of Drinking Water by Lead." *British Medical Journal*, August 31, pp. 459–462.

Whitsell, F. M. 1941. "Lead Pipe for Connection Between Water Main and Residence." *Journal of the American Medical Association* 120:995.

Wibberly, D. G., A. K. Khera, J. H. Edwards, et al. 1977. "Lead Levels in Human Placentae from Normal and Malformed Births." *Journal of Medical Genetics* 14:339–345.

Wilson, A. L. 1979. "Trace Metals in Water." *Philosophical Transactions of the Royal Socity of London. Series B, Biological Sciences* 288:25–39.

Wilson, Allen T. 1966. "Lead Absorption and the Health of a Community." *Practitioner* 197:77–85.

Winder, C. 1993. "Lead, Reproduction, and Development." *Neurotoxicology* 14:303–317.

Wrangham, W. 1901. "Acute Lead Poisoning in Women Resulting from the Use of Diachylon [lead plaster] as an Abortifacient." *British Medical Journal*, pp. 72–74.

Wright, Wade, Clarence O. Sappington, and Eleanor Rantoul. 1928. "Lead Poisoning from Lead Piped Water Supplies." *Journal of Industrial Hygiene* 10:234–252.

Wrigley, E. A., and R. S. Schofield. 1989. *A Population History of England, 1541–1871: A Reconstruction.* Cambridge: Cambridge University Press.

Wynne, F. E. 1911. "Domestic Hot-Water Supplies as a Factor in the Production of Lead Poisoning." *Lancet*, August 5, p. 385.

Xintaras, C. 1992. *Impact of Lead-Contaminated Soil on Public Health: An Analysis Paper.* Atlanta, Ga.: U.S. Department of Health and Human Services, Agency for Toxic Substances and Disease Registry. Available online at http:www .atsdr.cdc.gov/exlead.html.

Yiin, Lih-Ming, George C. Rhoads, and Paul J. Lioy. 2000. "Seasonal Influences on Childhood Lead Exposure." *Environmental Health Perspectives* 108:177–182.

Yoshida, Takahiko, Hiroshi Yamauchi, and Gui Fan Sun. 2004. "Chronic Health Effects in People Exposed to Arsenic Via the Drinking Water: Dose-Response Relationships in Review." *Toxicology and Applied Pharmacology* 198:243–252.

Young, James. 1914. "The Aetiology of Eclampsia and Albuminuria and Their Relation to Accidental Haemorrhage." *Journal of Obstetrics and Gynaecology of the British Empire* 26:1–28.

Young, James, and D. A. Miller. 1920. "Further Observations on the Cause of Eclampsia and the Pre-eclamptic State." *Medical Press*, December 15, p. 477.

Young, James, Jessie C. B. Sym, and Elsie V. Crowe. 1932. "An Evaluation of the Incidence of and the Maternal Disability following Eclampsia and Albuminuria." *Proceedings of the Royal Society of Medicine* 25:1235–1241.

Zietz, B., J. D. de Vergara, S. Kevekordes, et al. 2001. "Lead Contamination in Tap Water of Households with Children in Lower Saxony." *Science of the Total Environment* 275:19–26.

Index

abandonment of, 6, 202–204

adult health used to assess safety of, 16, 17, 62, 72, 75, 110, 115–116, 123, 149, 201

age of *(see* Water lead, and age of lead pipes)

in the ancient world, 63, 150–151, 158

consumer denial of risks of, 165–166

consumer misinformation regarding safety of, 18, 141, 150–162, 166–168

consumer preference for, 18, 83, 141

cost of, 18, 83, 135, 137, 139–140, 151, 197, 201

definition and description of, 10

durability of, 18, 151–152, 201–202

and engineers, 123, 136, 138, 151–152, 162, 189, 202 (*see also* *Engineering News*; New England Water Works Association; Thomas, R. J.)

frequent use of, 10–12, 138–140, 167, 201

length of, 10, 25, 166, 257n4

local ordinances mandating use of, 18, 141–144, 152, 164

more frequent in large cities, 10

and ownership of local water system (public or private), 197–200

physician opposition to, 6, 163–164, 196–197 (*see also* Adams, Horatio; Jones, Robin Russell; Kingsbury, George; Porritt, Norman; Swann, Alfred)

physician support of, 7–8 (*see also* Reese, Meredith)

plumber support of (*see* Gallagher, Joseph P.)

safety as situation-specific, 13, 14, 16, 22, 201

in specific places (*see* Boston, Chicago, etc.)

substitutes for, 6, 14, 137, 140, 143, 146–147, 152, 164, 201

Lime dosing. *See* Water treatment

Lindsay, Lauder A., 104, 129

Linenthal, Harry A., 109–111

Loch Katrine, 180. *See also* Glasgow; Loch Katrine aqueduct

beauty of, 169

lead solvency of water in, 169, 184–190, 194–197, 202

proximity of, to Glasgow, 169, 179

purity of, water, 169, 179, 181, 182, 185, 194

Loch Katrine aqueduct, 180. *See also* Glasgow; Loch Katrine

effect of, on mortality in Glasgow, 169, 182, 193, 194

complexity of, 181, 191, 194

construction of, 181

cost of, 169, 181, 192

as a model of municipal socialism, 169, 191, 202

promoters of, 178, 179, 180

London, 35, 84, 133, 176

cost of waterworks in, 192

safety of lead water pipes in, 13, 14, 124, 125

rarity of eclampsia in, 80, 89–90

use of lead water pipes in, 10

water hardness in, 89, 125

London doctrine. *See* Doctrine of protective power

Lowell (Massachusetts)

abandonment of lead water pipes in, 161–162

decision to install lead water pipes, 163

early history of water-related lead poisoning in, 163–164

epidemic of water-related lead poisoning in, 160–162

and Massachusetts State Board of Health, 160–162

water-lead levels in, 54–58, 160, 161

Lowell Daily Sun, 158, 159, 160

Magnesium, 38, 41, 45, 59, 124, 206

Maine, 130

Plumbosolvency. *See* Doctrine of protective power; Water hardness and lead solvency; Water lead, determinants of
Putnam, James J., 113–114

Rantoul, Eleanor, 114–117
Reese, Meredith, 8, 154–158

Sappington, Clarence O., 114–117
Scotland, 49, 80, 186, 187, 193, 196, 204, 257n4. *See also* Aberdeen (Scotland); Edinburgh; Glasgow (Scotland)
Scott, Sir Walter, 169, 182
Sheffield (England), 10, 77–78, 105, 137, 147, 187
epidemic of water-related lead poisoning in, 162–163
Smallpox, 69
Soda fountains. *See* Lead, and soda fountains
Spontaneous abortions. *See also* Infant mortality; Stillbirths
and lead exposure, 30, 37, 38, 66, 67, 69
and water lead, 36, 62–65, 72, 113, 120, 121
Stainthorpe, W. W., 64–65, 108–109
Steckel, Richard, 206
Stillbirths, 15–16, 30, 66, 67
and lead exposure, 36–38, 65
and water lead, 62, 64, 69, 75, 113, 159, 209–229
Swann, Alfred, 19, 23, 36, 62–65, 75, 201

Theobald, G. W., 80–81, 89
Thomas, R. J., 161–164
Thomson, William, 104–105
Thresh, John C., 72, 129, 132, 190
Tin-lined pipes, 137, 146–147, 152–155
Typhoid fever, 6, 69, 112–113, 138, 193, 197

Urine lead, 84, 99–101, 114
as an unreliable indicator of lead poisoning, 107

Vitruvius, 63, 158

Wales, 65, 78, 84, 86–93, 159, 187, 196
Washington, D.C., 13, 191, 205, 206
Water hardness. *See also* Infant mortality and
and cardiovascular disease, 206
and incidence of lead water pipes, 138–140, 201–202
and lead solvency, 17, 56, 59, 83, 84, 86–87, 124–125, 129–132, 135, 136, 140, 201–202, 251–253 (*see also* doctrine of protective power; Loch Katrine, and lead solvency of water; Water lead, determinants of)
Water lead. *See also* Lead; Lead poisoning; Lead water pipes
and acid rain, 127, 205
and age of lead pipes, 17, 59–60, 117, 123–125, 195, 202
and alkalinity, 131–133, 272n22, 272n23
in the ancient world, 63, 150–151, 158
awareness of, 16, 99–105, 120–121
and blood lead, 47–50, 195–196
as a cause of disease and pathology (*see* Eclampsia; Infant mortality; Lead poisoning, cases of water-related; Spontaneous abortions; Stillbirths; Water lead, and convulsions; Water lead, and infertility)
and changes over time, 6, 202–204
and CO_2 levels, 56, 133–136
and convulsions, 93–94, 118, 160, 248–250
current legal thresholds for, 5, 14, 195–196
current levels of, 5, 6, 195–196